Science Links

MW01680165

TIME TRAILS

A Study of the History of Life on Earth

Developed by
The Agency for Instructional Technology

AIT
The Learning Source

JOIN US ON THE INTERNET
WWW: http://www.thomson.com
EMAIL: findit@kiosk.thomson.com
A service of I(T)P®

South-Western Educational Publishing
an International Thomson Publishing company I(T)P®

Cincinnati • Albany, NY • Belmont, CA • Bonn • Boston • Detroit • Johannesburg • London • Madrid
Melbourne • Mexico City • New York • Paris • Singapore • Tokyo • Toronto • Washington

ISBN: 0-538-66877-6

3 4 5 6 7 8 PR 02 01 00 99

Printed in the United States of America

I(T)P®

International Thomson Publishing

South-Western Educational Publishing is a division of International Thomson Publishing, Inc.

The ITP trademark is used under license.

This book is printed on recycled, acid-free paper that meets Environmental Protection Agency standards.

SCIENCE LINKS STAFF

Project Design

Chief Consultant
Marvin Druger, Ph.D.
Department of Science Teaching and Department of Biology
Syracuse University

Instructional Designer
Diana W. Lee, Ph.D.
Agency for Instructional Technology

Project Developer
Jonathan Greenberg, Ph.D.
Agency for Instructional Technology

Project Management

Agency for Instructional Technology

Director of Projects and New Products
Frank J. Batavick

Project Management Specialist
Diane Sumner

Formative Evaluation
Rockman et al.

Administrative Assistant
Connie Williamson

South-Western

Publisher
Thomas A. Emrick

Project Manager
Karen Roberts

Editor
Marianne Miller

Marketing Manager
Susan Denney

Marketing Assistant
Kris M. White

Science Links Reviewers

Gary Abbas, West High School, Davenport, Iowa; **Robert Allen**, Victor Valley Community College, Victorville, California; **Carol Bedford**, Smart Intermediate School, Davenport, Iowa; **Jennifer Braught**, Shakamak High School, Jasonville, Indiana; **Matt Braught**, Linton-Stockton High School, Linton, Indiana; **Glenda Burrus**, Pinetown High School, Pinetown, North Carolina; **Laine Gurley-Dìlger**, Rolling Meadows High School, Rolling Meadows, Illinois; **Ron Endris**, Floyd Central High School, Floyds Knobs, Indiana; **Sheila Engel**, Smart Intermediate School, Davenport, Iowa; **Tom Ervin**, North High School, Davenport, Iowa; **Marcy Gaston**, Pike High School, Indianapolis, Indiana; **George Hague**, St. Mark's School of Texas, Dallas, Texas; **Mary Halsall**, Hughes Center, Cincinnati, Ohio; **Craig Leonard**, Abraham Lincoln High School, San Francisco, California; **Ann Lumsden**, Florida State University, Tallahassee, Florida; **Jim Oberdorf**, Lincoln High School, San Francisco, California; **Jan Pierson**, Bloomington North High School, Bloomington, Indiana; **Eva M. Rambo**, Bloomington South High School, Bloomington, Indiana; **Willa Ramsay**, Madison High School, San Diego, California; **Kathleen Schuehler**, Liverpool High School, Liverpool, New York; **Len Sharp**, Liverpool High School, Liverpool, New York; **Dwight Sieggreen**, Cooke Middle School, Northville, Michigan; **Gerald Skoog**, Texas Tech University, Lubbock, Texas; **Ernestine B. Smith**, Tarboro High School, Tarboro, North Carolina; **Charlotte St. Romain**, Carenco High School, Lafayette, Louisiana; **Denise Tompkins**, Southwest Edgecomb High School, Pine Tops, North Carolina; **Kevin Tsui**, Woodside High School, Woodside, California; **Rick Wells**, Central High School, Davenport, Iowa; **Eric Worch**, Indiana University, Bloomington, Indiana

Video and Print Development

Video

Executive Producer
David Gudaitis, Ph.D.

Associate Producer
Brad Bloom

Video Editors
Amy Crowell
John MacGibbon

Print

Senior Editor
John Pesta

Features Writer/Editor
Lisa Williams

Safety Editor
Doug Mandt

Science Writers
Sara Brudnoy, Ph.D.
Robert Fariel
Lillian Fitzpatrick
Jonathan Greenberg, Ph.D.
Judy E. Hammett
Natasha X. Jacobs
Michael Svec, Ph.D.
Laura Twardy

Assessment Writers
Jennifer L. Chidsey
Laura Henriques
Margaret A. Jorgensen, Ph.D.
James A. Shymansky, Ph.D.

Layout Designer/Compositor
David Strange

Cover and Page Designer
Brenda Grannan

Print and Art Coordinator
Jay Hagenow

CONTENTS

Science Links . . . Where You and Your Future Connect

You can't see them, but there are thousands of doorways in front of you . . . doors to your left, doors to your right . . . big doors, little doors, revolving doors, trapdoors…doors stretching into the distance as far as you can see. . . . Each of these doorways is a possibility, an opportunity for a successful and prosperous life. But most of the doors are locked. You need the right keys to open them. How many doors will you be able to open?

The more you learn in school, the more keys you will have . . . the more choices . . . the more control over your own life.

Science Links is one key to your future. In the world of tomorrow—the world of computers, robotics, the World Wide Web, gene therapy, space travel—your knowledge of science can help you keep up with the latest developments in technology. More important, it can help you find a good job and pursue a great career.

Many students consider science a difficult subject, but as you will see in this course, science can be fascinating. In ***Science Links*** you will learn science by doing many laboratory investigations. (Investigations that are too dangerous for the classroom will be demonstrated in video segments.) These experiments will help you understand scientific principles.

Your Link to Science

Science is an attempt to make logical sense out of the natural world. It generally involves observations, experiments, data collection, analyses, and logical conclusions. ***Science Links*** will let you experience science, and the experience will be informative, interesting, and fun. You will be doing many laboratory investigations that will demonstrate the importance of science in your life, both now and in your future career. Following one or more investigations, a brief reading will explain the scientific principles further.

This course gives you a great opportunity to learn about science, but the learning is up to you. Learning science can be challenging, but if you work at it, you can master the contents and processes of this fascinating field. Put your full effort into this course and you will gain the satisfaction of learning many useful things about science. More important, you will advance to the next level of your education with greater confidence, a better self-image, and a strong motivation to learn even more.

So work hard, and enjoy your ***Science Links*** experiences.

—Professor Marvin Druger, Past President
National Science Teachers Association

Professor Druger teaches biology at Syracuse University, where he is also chairperson of the Department of Science Teaching.

You will perform most of your investigations as a member of a small group. After your group completes its work, you and your teammates will have an opportunity to present your findings to the other students in your class. In the real world—the world of work—many jobs are based on teamwork, men and women working together to solve problems and share information.

In a few years you will be joining the adult workforce. ***Science Links*** can help prepare you for that important step in your life.

To give you a broad introduction to science, ***Science Links*** blends biology, chemistry, physics, and earth/space science into a single course. You will learn how various scientific concepts are used in the everyday world, both at home and in the workplace.

This book and the related video segments will also show you how science is used in dozens of careers, from food science to auto mechanics, from welding to respiratory therapy. Check out these careers. Maybe one of them is right for you.

When you picked up this book, you picked up a golden key. Use it now. Open the ***Science Links*** door, and step right in to the amazing world of science.

Log It!

As you perform the investigations in ***Science Links***, you will record various kinds of information in a logbook. Use your logbook to take notes, to collect data, and to enter your observations and conclusions about the experiments you perform. The logbook is also a great place to jot down any questions and ideas that occur to you, either inside or outside the classroom. Put them on paper before you forget them!

If you get in the habit of using your logbook, it will help you learn to express your thoughts simply and clearly—that's good writing. The logbook will also give you a permanent record of your work—that's good science.

For more suggestions on using your logbook, see Appendix A (pages 80–81). The appendix also contains a sample logbook page.

Safety Symbols

TAKE appropriate precautions whenever these safety symbols appear at the beginning of the Step-by-Step instructions. All safety icons that apply to a particular investigation appear at the beginning of that investigation. In addition, a step number that is printed in **red** indicates the first time a certain kind of safety hazard exists in an investigation.

Disposal Hazard

- Dispose of this chemical only as directed.

Fire Hazard

- Tie back hair and loose clothing.
- Do not use a burner or flame near flammable materials.

Eye Hazard

- Always wear safety goggles.

Poison Hazard

- Do not chew gum, drink, or eat in the lab.
- Keep your hands and all chemicals away from your face.

Inhalation Hazard

- Avoid inhaling the substance.

Thermal-Burn Hazard

- Wear gloves and do not touch hot equipment.

Breakage Hazard

- Do not use chipped or cracked glassware.
- Do not heat the bottom of a test tube.

Corrosive-Substance Hazard

- Wear safety goggles and a lab apron.
- Do not touch any chemical.

IN CASE OF EMERGENCY . . .

Immediately report any accident, injury, or spill to your teacher. Know where to find the nearest fire blanket, fire extinguisher, eyewash, sink, and shower.

Here's what to do:

Fire—Turn off all gas outlets and unplug all appliances. Use a fire blanket or fire extinguisher to smother the flames. When using a fire blanket or extinguisher to smother flames, take care not to cut off or impede the victim's air supply.

Burn—Flush the affected part with cold water.

Poisoning—Take note of the substance involved and call the teacher immediately.

Eye Injury—Flush eyes with running water; remove contact lenses. Do not allow injured persons to rub their eyes if a foreign substance is present.

Fainting—Open a window or provide fresh air as best you can. Position the person so the head is lower than the rest of the body. If breathing stops, use artificial respiration.

Spill on Skin—Flush with water.

Minor Cut—Allow to bleed briefly and then wash with soap and water. If necessary, apply a bandage.

Remember to call your teacher right away in any emergency!

module 8

Time Trails

Section 1

THE RULES OF THE GAME

All fish live in water. But a shark can live only in salt water, while a bluegill needs fresh water. What determines where different kinds of animals live?

Why is it that some members of a species live longer than others? One deer may survive for many years, producing offspring again and again. But another deer may survive just long enough to become a meal for coyotes.

In the opening video segment, you will see animals and plants living in all sorts of places.

What makes life change?

In Section 1 you will discover how living things survive in different environments. Living things can be found almost everywhere. But what happens to them when their environment changes? What happened to the dinosaurs that once roamed this planet? You will discover the answers to these questions as you explore the history of life on earth in Sections 2 and 3. Then you will look into the future in Section 4.

Freeze Frame

Now that you have watched the video segment, discuss the following on-screen questions with the other students in your class.

1. Would you find a greater variety of plants and animals in northern Alaska or in Central America?
2. People can generally control their environment. How do plants and animals cope with environmental conditions?

INVESTIGATION 1

The Name of the Game Is Survival

Ever Wonder?

Many people like to plant flowers and vegetables in the spring. Maybe you have helped plant a garden in your own backyard. If you have, you probably noticed that not every seed or seedling grew. Some of them died. Did you wonder why some grew and some did not? Did you think you did something wrong? In this investigation, the name of the game is survival. You will explore why some members of a species survive and succeed while others do not.

Materials List

- Tokens representing worms (two colors)
- Plastic or paper bags
- Pocket calculator

Step-by-Step

1. Scatter 50 "worms" of each color in the search area that your teacher has assigned to your group. Be sure you stay within your designated area, and do not attempt to hide the worms.
2. Look at your group's copy of the data form that your teacher has provided. Have one member of your group serve as recorder.

3. Now examine your search area. The recorder should list the color of this area on the blank line at the top of the data form. The recorder should also list the colors of your worms at the top of the third and fourth columns. Then the recorder should complete line 1 in the **Generation 1** section of the form.

4. The other member(s) of your group will search for worms. When your teacher gives the signal, the searcher(s) should pick up as many worms as possible using only one hand. Place the worms in a bag. Continue picking up worms until your teacher says to stop.

5. Count how many worms of each color your group collected. The recorder should then fill in line 2 of the data form.

6. Work as a group to do the calculations shown on lines 3–6 of the data form. The recorder should then complete these lines. Report your results to your teacher.

7. You will now use the "worms" that are still on the ground to represent part of a new generation. You will need to bring the total number of worms in this generation up to 100. To do this, scatter the number of worms of each color that the recorder listed on line 6 of your data form.

8. Have the recorder complete line 1 of the **Generation 2** section of your data form.

9. Repeat the procedure you followed in Steps 4–6, but this time have the recorder complete the **Generation 2** section of the data form.

10. Repeat the procedure two more times, completing the **Generation 3** and **Generation 4** sections of your data form.

11. Collect all your worms. Store them according to your teacher's instructions.

12. When your teacher has completed the table summarizing the whole class's data, copy it in your logbook.

Talk It Over

Work with the other member(s) of your group to answer the following questions.

1. How did the number of worms of each color change over four generations? Did the number of each color change the same way in all groups? How would you explain any differences among the groups?

2. What caused the change in the number of worms of each color? Could the same kind of change occur with real worms living in a field? Explain your ideas.

3. What do you think would happen to real worms if they all were the same color? How did your worm population benefit from having two different colors?

4. How could individual differences in color benefit snails living in a wild meadow?

5. What do you think would happen to the colors of your worm population after another 10 generations?

Spread the Word

Develop a presentation so that the rest of the class can share what you learned in this investigation. One way to begin would be to make a poster showing differently colored organisms on different backgrounds. You could explain how the poster illustrates the relationship between environment and an organism's ability to survive. Tell the class how this information can help people understand the importance of differences in the physical characteristics of plants and animals. Use specific examples, such as wild rabbits with different speeds or wolves with light or dark fur.

Tell how farmers can use your findings. Then describe at least one other career or situation that requires an understanding of the connection between an environment and the individual differences within a population.

Read All about It!

Competing to Survive

No one would try to keep a goldfish in a hamster cage—or a hamster in an aquarium filled with water. The obvious differences between a hamster and a goldfish make it impossible for them to live in the same environment. Differences also exist between one goldfish and another. Those differences are much less obvious, but they are just as important when it comes to survival.

If you look at two goldfish, you will see that they are not exactly alike. One may be a different shade of gold than the other, or they may even be completely different colors. A difference in the traits of closely related individuals is a **variation**. The "worms" that you examined in Investigation 1 showed some variations in color. You can see plenty of variation by looking around your classroom. Everyone in your class has a different height, weight, and coloring. Some people have excellent vision, while others need to wear glasses. Despite individual differences, all your classmates are members of the same **species**, just as all goldfish, all hamsters, and all corn plants belong to their own species.

What It Means to Belong

"Belonging to the same species" means that individual members can interbreed with one

These happy-face spiders show the great variation that exists within this species. Because no type is better adapted than the others, many types survive. Variation makes the population more adaptable.

another and produce fertile offspring—a new generation that will also be able to reproduce. Every offspring inherits some characteristics from each of its parents. As a result, every individual has a unique combination of traits and may show even more variation than its parents.

Although some species vary more than others, even slight differences among the individuals within a species can have major consequences. For example, trout live in both flowing water, such as rivers and brooks, and in still water, such as lakes. If a number of brook trout became trapped in a pond, minor differences in body chemistry would be critical. The trout that could tolerate the lower oxygen level of still water would have the best chance of surviving and reproducing. Trout that need the extra oxygen dissolved in moving, bubbling creek water would be in serious trouble.

It is easy to see which kind of trout would be more likely to survive and reproduce as the fish compete for the limited amount of food and oxygen in the pond. This is an example of **competition**.

And the Winner Is . . .

Competition does not have to involve fighting or some other kind of direct contact. Individuals also compete simply on the basis of their traits. Think about the so-called worms in your investigation. They were competing with one another.

The Benefits of Blending In

Is it safer to be a light-colored moth or a dark-colored moth? If the peppered moth could talk, it might say, "It all depends on the time and place."

The peppered moth is a good example of how one variation or another can become the most common type, depending on the environment. Ever since this species appeared, there have been light-winged and dark-winged varieties. Until 200 years ago, the light moths were the dominant variety. They had become dominant because both types liked to perch on gray lichen that grew on the trunks of beech trees, which have whitish bark. The light moths blended in with this background, making it hard for predators to spot them. But the dark moths showed up against the lichen and tree bark. As a result, the dark moths tended to be eaten before they could reproduce.

The industrial revolution, which began in the late eighteenth century, changed the peppered moth's world. Smokestacks filled the air with soot. The soot often killed the lichen and turned the trunks of beech trees black. By the end of the nineteenth century, the gray moths had become a tiny minority. Now they were the ones that stood out against the trees and were easy prey.

The light moths did not die out. Enough survived to carry on their variety—just as the dark ones had earlier. During the twentieth century, many laws were passed to reduce pollution and clean up the environment. Thanks to these changes, the light-winged peppered moth has been making a comeback.

Those that were not picked up "won" the competition: They survived to produce the next generation. Real worms whose color makes them harder to see on the ground are more likely to escape being eaten by birds. With each new generation, worms of this color will become a larger percentage of the population.

What if every worm were the same color—or if every brook trout had exactly the same body chemistry? If no variation existed in the population, its members would be able to survive only in a small range of environments. Brook trout could survive only in streams, and your worms could survive only in one type of search area. Natural variation gives the members of a species a chance to survive in a new or changing environment.

The Pressure's On

What is true for trout is true for all species of living things. Variation among the individual members of a species makes it possible for that species to survive if its environment changes. However, the variation that occurs to a particular member of a population may be either helpful or harmful to that individual. For example, if a green grasshopper hatched out of an egg on a grassy lawn, it would have a better chance of surviving than a yellow grasshopper would. As you saw during the investigation, the environment serves as a testing ground for every individual's set of characteristics. Every environment "selects" some types of individuals over others. In other words, the environment causes a **selection pressure** on the population.

Variation among the individual members of a species makes it possible for that species to survive if its environment changes.

Individuals with characteristics that help them to compete against other individuals have a better chance to survive and reproduce. The ones that are at a disadvantage are less likely to survive. They will not live as long; therefore, they will produce fewer offspring. Gradually, over many generations, the more successful types of individuals will become the majority of the population. The species will **adapt** to its environment—it will become better able to survive there.

The anteater is a dramatic example of adaptation. Many generations of competition and natural selection have produced an animal well suited to eating insects.

Nature's Way

The process by which some members of a species—those with certain kinds of traits—survive and reproduce, while others do not, is called **natural selection**. The worms, trout, eagles, pine trees, and other organisms that have the traits that allow

Which Way to Go—High or Low?

The most famous animal that Charles Darwin encountered on his voyage to the Galápagos Islands is the giant tortoise. He found 14 varieties of this species. Each was so distinct that an official of the islands told Darwin he could look at any tortoise and tell which island was its home.

One difference among tortoises is the shell. Some large tortoises have domed shells (top photo). Others, which are called saddlebacks, have shells that look raised in front—like saddles (bottom photo). Domed tortoises live in the higher, wetter islands, where they feed on fallen fruit, grasses, and other surface plants. Saddlebacks live in lower environments that are hot and dry and where the vegetation is not easily accessible.

The saddleback's raised shell allows it to stretch out its long neck to reach its favorite food, the prickly-pear cactus. Prickly-pears on those islands are tall and treelike, and they have stiff, sharp spines when they are young. Only the tortoise that can stretch its neck can reach the edible parts of these plants. And only the tall and spiny prickly-pear is found on islands that have saddlebacks. A soft variety of this cactus also exists in the Galápagos, but it grows only on islands with no tortoises.

Darwin's observations of tortoises gave him insight into variation and natural selection. The type of tortoise that would succeed on a particular island was the one best adapted to its environment.

them to survive for the longest time and to reproduce the most offspring are the ones most likely to succeed. That's bad news for individuals with other traits, but the process helps the species as a whole to survive by better adapting to its environment. Basically, natural selection involves the ability of a species to pass on inheritable traits from one generation to the next. As the process of natural selection plays out over many generations, those individuals that do not have the traits best suited for long-term survival will gradually decline in number. But the species as a whole will benefit from the process: The percentage of individuals that *do* have the necessary traits will gradually increase.

Get This . . .

Competition does not have to involve fighting. In fact, individuals can compete for food, living space, or other resources without making direct contact with one another.

On Your Own

Answer the following questions based on "Competing to Survive."

1. Many people associate natural selection with survival of the biggest, strongest, and fastest. Is this an accurate idea of the process of natural selection? What are some situations in which natural selection probably does favor bigger, stronger, faster animals? What are some situations in which the opposite may be true? What are some situations in which other types of traits affect which individuals survive and reproduce?
2. Plants do not move around looking for food and mates. What are some ways that plants compete with one another?
3. How could competition between two tigers for the same territory benefit future generations of tigers?

Variety—More Than Spice

Variety has been called the spice of life. It may be more than that. A lack of variety may mean the end of a species.

The Eltham copper butterfly is a vulnerable species that lives in Australia. It is not endangered by predators, pollution, or even human collectors, as some scarce butterfly species are. The Eltham copper is at risk because of changes in its environment and its own lack of variation.

This butterfly lays its eggs only on one type of plant, *Bursaria spinosa*, which is its only source of food. Furthermore, the Eltham copper needs this plant for more than food. At the base of these plants lives a certain type of ant. For some reason, possibly an attraction to smell, these ants carry the caterpillars of the Eltham copper butterfly below ground to their nests, where they are kept safe.

Human population growth has had a negative effect on *Bursaria spinosa*. As the plant's numbers have dwindled, so has the population of the Eltham copper butterfly.

INVESTIGATION 2

One Big Family

Ever Wonder?

You probably learned a long time ago that cats, lions, tigers, and leopards are all part of the "cat family." Maybe you have heard of the "dog family" and some other kinds of animal families. What about all the different kinds of snakes? Are they all one big family? How about different kinds of palm trees—are they all related too? What does it mean to say that two different species, such as a house cat and a tiger, are related to each other? And how did there get to be so much variety within a single family of animals? You will begin discovering the answers to these questions as you carry out this investigation.

Materials List

- 100 pieces of "food" (50 sticky and 50 nonsticky) per student
- Bags or bowls
- Masking tape
- Ice-cream sticks

Step-by-Step

Generation 1

1 Your teacher will provide a supply of "food." Scatter it in the search area according to your teacher's directions.

2 Imagine that two types of hands occur in the population of a certain species. Your teacher will assign you either type A or type B.

3 Look at Figure 1. Tape your right hand (or your left, if you are left-handed) according to the type you have been assigned.

4 When your teacher gives the signal, start foraging (searching) for food. Hold your bag in your free hand. Using only your taped hand, pick up as many pieces of food as you can and put them into the bag. When your teacher says time is up, immediately stop foraging.

5 You have collected two different kinds of food: sticky and nonsticky. Count the number of pieces of each type in your bag.

6 Add up the calories. Each piece of nonsticky food contains two calories. Each piece of sticky food contains one calorie.

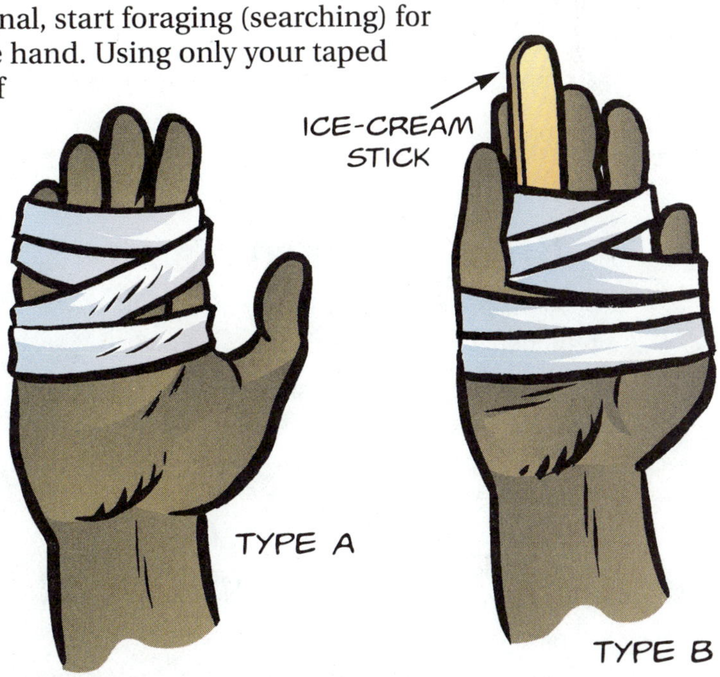

Figure 1: Hands prepared for Investigation 2

7 Report your type of hand, the number of sticky and nonsticky food pieces that you collected, and your total number of calories to your teacher.

8 Return the food pieces that you collected to your teacher.

9 Your teacher will tell you how many calories you needed to collect in order to "survive." If you collected that many calories or more, raise your hand to let your teacher know that you are a survivor.

10 If you are not a survivor, move away from the foraging area and untape your hand. Pay attention to what happens in the next generation.

Generation 2

11 If you are a survivor, you will participate in Generation 2. Repeat Steps 4–10. Wait for your teacher's signal to start foraging.

Generation 3

12 If you survive Generation 2, you will take part in Generation 3. Repeat Steps 4–10. Again wait for the signal to start searching for food.

Generation 4

13 If you survive Generation 3, you will participate in Generation 4. Repeat Steps 4–10. Start foraging when you get the signal.

All Students

14 Pick up the remaining pieces of food. Return all pieces to your teacher.

15 In your logbook, copy all the data that your teacher has written on the chalkboard.

16 Form discussion groups, according to your teacher's instructions.

Beautiful Death Traps

Animals eat plants. So why shouldn't plants eat animals?

Some of them do. Most animal-eating plants are insectivores—insect eaters—but some can catch and digest vertebrate animals up to the size of small rodents.

Most plants receive enough of their nutrients from the soil. Animal-eating plants developed as an adaptation to soils that are poor in nitrogen.

The best-known bug-eating plant is the Venus flytrap, which is native to the United States. The flytrap's method of capture is unique: It has hinged leaves with small hairs on them. If a fly or some other small insect touches one of the hairs, it triggers the hinge to snap the leaf shut. This traps the bug inside the leaf, where the plant digests it.

The North American bladderwort has a different style of catching its prey. It is chiefly an aquatic plant, with its flowers growing above water and its traps, or bladders, below. A swimming creature may accidentally open one of the plant's "trap-doors," allowing water to rush into the empty trap and take the swimmer in with it. Then the door shuts again. This takes about 1/30th of a second, giving the prey no time to escape.

The bladderwort has a carnivorous relative, the bloodwort. Its specialized leaves roll up and then digest the prey.

The largest and most diversified of the carnivorous plants are the Nepenthes, or pitcher plants. These vines live in tropical regions. They sometimes grow nearly 20 meters long and can climb trees. The pitchers attract insects, which fly into them and get trapped. The Nepenthes can also trap vertebrates, such as birds, rodents, and amphibians.

Talk It Over

Work with the other members of your group to come up with answers for the following questions.

1. Did students with both types of hands look for food in the same area, or did they separate to search in different places? Did one type prefer the sticky food? If so, how would you explain this?

2. Suppose the sticky food were poisonous to most individuals with type-A hands. Would you expect this to cause more or less variation among individuals who eat the sticky food?

3. Imagine that the two kinds of food are located at different ends of a very long beach and that each type is eaten by a different type of crab. Would you expect to find many crabs that eat sticky food living at the end of the beach where the nonsticky food is? Give your reasons.

4. Would you expect the crabs that eat sticky food to go to the other end of the beach in search of mates? What do you think would happen to the populations of the two kinds of crabs after many generations?

5. Would the crabs experience more competition if the two groups remained separate or if they lived in the same place? Explain their separation based on what you know about competition.

Career Links

Agriculture and Natural Resources

U.S. Forest Service

Horticulturist

Horticulturists breed, grow, and care for plants. Many work in plant nurseries or greenhouses. Some specialize in landscaping for homes, businesses, parks, golf courses, and other sites. Different plants have different requirements for soil type, planting depth, water, light, and fertilization. Horticulturists must know about these various needs. They also must be able to identify and control plant diseases and pests. Horticulturists who specialize in developing new breeds of plants need an understanding of genetics.

To pursue a career in this field, you will need a two-year associate's degree in applied science or a four-year degree in horticulture. For more information, write:

The American Society for Horticultural Science
600 Cameron Street
Alexandria, VA 22314
http://www.ashs.org

Related occupations: botanist, agribusiness consultant, groundskeeper, plant pathologist, florist

Spread the Word

Work with the other students in your group to share your conclusions with the rest of the class. To compare what happened with the two types of hands, you could create a graph or table on a poster. Explain what caused the population to separate into two different groups with different physical characteristics. Then describe other situations in which a group of individuals might split off from an original population.

Read All about It!

Something Special Happens When a Species Specializes

How would you like to work in the recording industry and produce CDs with your favorite music? It's a tough business. With so many performers, there is a lot of competition out there. How could you compete against the major companies, with all their money and clout? How could you hope to stay in business? One way would be to specialize in some type of music that most people never hear—maybe the latest sound from Brazil or Gaelic folk ballads or a new style of rap. By carving out your own niche, you would face less competition as you worked to develop a devoted group of customers.

Many species of plants and animals deal with competition in a similar way. For example, there are thousands of species of insects, far more species than any other kind of animal. If all those species had to compete for the same kinds of food and the same places to live, very few of them would be able to survive. But different species specialize in different things. They eat different foods and live in different habitats.

Flopsy, Mopsy, and Lichen

When you were little, you may have heard the tales of Peter Rabbit. Peter and his friends were created by Beatrix Potter, a famous writer and illustrator who produced her books in the early part of the twentieth century. While many people know Potter's stories, probably few know that she was also a scientist.

Potter studied lichen. Lichen is an example of symbiosis (two or more species living together for mutual benefit). A lichen looks like a single plant, but it actually is composed of one or more types of algae growing in close association with a fungus. Algal cells are woven into a mass of fungi. Many species of algae that form a symbiotic relationship with fungi are not found living separately. They may need their "marriage" with fungi to survive.

In Potter's time, the dual nature of lichens was controversial. She accepted the symbiotic theory, and she wrote an article about it. The article was filled with her own drawings of what she had observed under the microscope. The paper was read before the Linnean Society, an important scientific organization—but not by Potter. In those days, women weren't allowed to attend the meetings.

How times have changed. The Linnean Society recently chose Potter's drawings of plants to illustrate a book promoting conservation.

It Could Happen to Anyone

How did there get to be so many different species of insects? Insects are not like recording-industry executives—bugs do not make conscious decisions to specialize in one thing or another. They become specialists through the natural process of heredity, among other things. The term heredity refers to

Mutation and natural selection have produced more than a million species of insects, each with unique adaptations. The leaf butterfly (right) is camouflaged and uses a coiled tube to sip nectar. The dead leaf mantid (top) has spiked front legs to hold prey. The jaws of the Atlas beetle aid in feeding and defense.

the inherited traits that one generation passes on to the next. As genetic information is copied within the cells of a parent, sometimes a "mistake" occurs. That mistake may cause a change, or **mutation**, that the offspring will inherit. In other words, the offspring will have a new trait the previous generation did not have. These mutations occur at random. Sometimes they are helpful, sometimes harmful. Often the changes are so minor that they have little effect on an organism's ability to survive and reproduce.

When mutations appear within a population, the process of natural selection begins to act on them. Individuals with harmful mutations are less well adapted than the "normal" members of their species. Because they are less likely to survive a long time, they are less likely to produce many offspring with the mutant trait. Such mutations may not last more than a few generations before the individuals that carry them die out.

On the other hand, when a mutation is beneficial, it tends to become more common with each generation. The reason is that individuals with the helpful mutation are more successful at surviving and reproducing than those that do not have such a mutation.

You Go Your Way, I'll Go Mine

Sometimes those individuals that inherit a mutation are more likely to mate with one another than with the rest of the population. For example, two similar species of squirrels live on opposite sides of the Grand Canyon in Arizona. Researchers think that long ago, before the Colorado River cut the canyon so deep, the squirrels may have been able to cross the river and mate with squirrels on the other side. Later, as the river widened and the canyon grew deeper, it became more difficult for members of the two populations to meet. They eventually found themselves in **geographic isolation**.

During many generations of isolation from each other, mutation and natural selection caused each group to adapt more closely to its own environment. Today the squirrels are different colors. Each type is well camouflaged in its own environment. Moreover, the two populations are no longer able to produce fertile offspring together—they have become separate species. The process by which a new species develops is called **speciation**.

Part of a population can break off from the original group and become a new species without a geographical barrier such as the Grand Canyon. For example, a certain type of fly lays its eggs on apple or hawthorn trees. When the eggs hatch, the maggots (fly larvas) eat the fruit on which the eggs were laid. At one time, the flies may have laid their eggs on either type of fruit; however, because of mutation, some of the flies now prefer apples while others prefer hawthorn. Although the flies live in the same places, each type tends to mate and lay eggs on its own favorite type of tree. As a result, the population is gradually separating into two species, even though no geographic barrier divides the two groups. The separation benefits all the flies because they experience less competition.

This kind of division into separate species, in which each species uses different resources in the same environment, is called **adaptive radiation**. Just like a recording company that specializes in one type of music, each species survives in part because it avoids competing with similar species.

GET THIS . . .

Mutations occur at random. Sometimes they are helpful, sometimes harmful.

WITH SEX OR WITHOUT

Who needs sex? Most plants and animals need sex to reproduce, but many other species do not.

There are two forms of reproduction: sexual and asexual. Asexual reproduction is very common. It's the way bacteria reproduce. They simply copy their own genetic material and divide. Even some vertebrates reproduce asexually. Some lizards and salamander eggs develop without being fertilized. This results in clones of the parent animal. Asexual reproduction has some advantages: It continues a successful genotype, and it eliminates the need to compete to find a mate.

Most vertebrates, invertebrates, and plants engage in sexual reproduction, which has its own advantages. Sexual competition often means that only the strongest, healthiest organisms mate. In addition, the two parents may share the work of protecting and raising their offspring.

Various plants and fungi practice both sexual and asexual reproduction. One animal that does this is the starfish. To reproduce sexually, an "arm" separates from the rest of its body. Under certain conditions, the arm will grow into a complete copy of the original animal.

ON YOUR OWN

Answer the following questions based on "Something Special Happens When a Species Specializes."

1. Suppose you found a dandelion that had a red flower instead of the usual yellow one. Could you call this a new species? Explain why or why not.

2. What are some ways that adaptive radiation could make it possible for several species of mice to live in the same forest? Explain each example.

3. In the past, sailors sometimes left cats and other animals behind on islands that ships rarely visited. The descendants of these cats have mated only with one another—they have had no contact with other cats. Does this mean these cats are a separate new species?

4. Giant pandas that live in the wild eat almost nothing but bamboo. Now and then all the bamboo in a wide area flowers and dies. If this happened every five years, would you expect the pandas to undergo adaptive radiation? If the bamboo died every 200 years, would the pandas experience adaptive radiation in that case? How would the pandas be affected if they found another type of food in the same area as the bamboo or in a different area? Explain your answers.

SECTION WRAP-UP

Life Is Full of Changes

EVERY spring billions of birds, bugs, and weeds produce far more offspring than can possibly survive. Every single one's chance of survival is affected by the environment in which it lives. Everything—a stray cat, a termite inspector, a speeding car, a flash flood—determines how difficult it will be for an individual organism to find a meal, a home, and a mate. Some individuals survive to produce offspring of their own. But many organisms do not survive long enough to reproduce. Variation gives some individuals an advantage over others. Those that are better adapted to their environment—for example, a moth that looks like tree bark—stand a better chance of surviving and reproducing.

Individuals that succeed in reproducing pass on to their offspring the traits that helped them to survive. The offspring are then subject to natural selection: They may or may not survive. It all depends on the environmental conditions they face.

Mutation causes many variations in individuals. Natural selection determines which individuals survive to pass on their variations.

It's a Test

The process of natural selection is a test of the individual's ability to adapt to environmental conditions. The test includes competition from members of one's own species. Competition for resources—food, shelter, mates—is common throughout the natural world. It is easy to see why this competition exists: The members of most species that reproduce sexually must live near one another to maintain a breeding population. If they all have the same needs and wants, there will be competition.

Many characteristics of plants and animals result from competition. A good example of this is the variation that exists among European bee orchids. Orchids depend on insects for fertilization. Each type of European bee orchid attracts a different type of bee. To do this, each type of orchid has specialized colors and markings. Orchids that do not attract the right bees cannot reproduce. Because each type of orchid attracts only one type of bee, there is less competition among the whole population of orchids; therefore, every orchid has a better chance of attracting the right kind of bee. (If all the orchids were competing for one kind of bee, the competition would be much greater, making it more likely that a particular orchid would attract no bees.) By specializing, the orchids—and the bees—reduce competition.

Adapt or Else

Competition can help to produce change in a population. Those individuals with the most favorable characteristics survive and mate. For example, in a cool climate where snails must warm themselves in the sun, dark-shelled snails would be better equipped to hide on the ground than white-shelled snails would be. But individuals with less suitable characteristics, such as the white snails, do not necessarily die or fail to reproduce. They might be more vulnerable, but some of them may still survive. And sometimes their traits may enable them to use other resources in

the environment. The process of adaptive radiation might permit white-shelled snails to survive in a sandy area, where their light color would help to protect them from hungry birds. If this were to happen, both types could survive and reproduce. However, if they must live in different environments, one type is less likely to mate with the other. Eventually both types may accumulate so many mutations that they would lose the ability to interbreed. They would become separate species.

The process of adaptive radiation can also be triggered by a geographical barrier. For example, members of the same species may become separated by a body of water. But whether or not a geographical barrier is involved, if two groups of a species live in reproductive isolation from each other, they will be unable to exchange genes. As a result, the two populations may diverge genetically and reach the point where they are no longer able to interbreed.

The Lottery of Life

Chance also plays a role in change. Mutations occur randomly. Through natural selection, individuals with helpful mutations tend to survive and reproduce, but individuals with harmful mutations tend to die off. For example, suppose a mutation occurs in a gene that helps to regulate how much sugar a plant produces. If this mutation were to reduce the plant's sugar production, it would make the plant less tasty to rabbits that like to eat it.

Such a mutation would be beneficial for the plant because more of them would survive—rabbits would eat the sweeter plants that did not have the mutant gene. Eventually most of the surviving plant population would carry the new version of the gene. (If no rabbits and other plant eaters lived in the area, the mutation would be less likely to become a large percentage of the population.)

Mutation causes many variations in individuals. Natural selection determines which individuals survive to pass on their variations. The more variations that exist within a population, the more likely the population is to branch out in new, diverse directions.

> **Get This . . .**
>
> **Natural selection** is the process by which some members of a species—those with certain kinds of traits—survive and reproduce, while others do not. The process helps the species as a whole to survive by better adapting to its environment.

On Your Own

Answer the following questions based on "Life Is Full of Changes."

1. What do you think would happen if all the offspring of every living thing survived? What does this have to do with natural selection?
2. Suppose an entire meadow were bulldozed for a new housing development. Would you say that each plant and grasshopper had unfavorable characteristics and died as a result of natural selection? Explain
3. Does natural selection cause a species to improve? Explain why it would or would not be correct to say that natural selection makes a species better.

Section 2

LIFE GETS A FOOTHOLD

If you understand natural selection, you understand how species of plants and animals can change in response to environmental conditions. But natural selection does not explain how life began. Did seaweed, worms, and other life-forms simply appear one day? How did the process of natural selection begin?

How did the first life on Earth look? Did many different kinds of organisms exist? Were the first ones anything like those that exist today? If not, how did they develop into today's organisms, and how long did that take? What kinds of changes in the environment led to changes in early organisms? Were there many "dead ends"? How could you learn about those dead ends?

In this section you will learn about the early history of life on the planet Earth. As you watch the next video segment, notice the environmental conditions that may have existed millions of years ago when life began. Think about what would have been needed for life to get started. Also consider how the process of natural selection may have affected the first forms of life.

Freeze Frame

Now that you have seen the video segment, discuss the following on-screen questions with the rest of your class.

1. What does evidence indicate the planet Earth was like when it formed? How has it changed?
2. How could the first living things have developed?

What did it take to bring the first living cells to life? Could it happen again? Could you make it happen right here in the classroom? Your teacher will now perform a demonstration. As you watch, think about what it would take to create a living thing from materials that are not alive. Then discuss what happens with the other students in the class. After your discussion, read "The Origin of Life."

READ ALL ABOUT IT!

The Origin of Life

How in the world did life gain a foothold on the third rock from the sun? No one knows for sure, but the stage was set an estimated 4.6 billion years ago when Earth formed. As the solar system took shape, a swirling cloud of dust and gases was drawn together by gravity to form this planet.

Over millions of years, this material became more and more compressed. Eventually it was so tightly compacted that great pressure and high temperatures built up. The new planet became a thick, boiling soup of molten rock.

It took another billion years or so for the surface to cool down enough to form a crust. Heavy minerals such as iron settled to the center, and lighter minerals rose to the surface. As Earth cooled, gases that were trapped inside the planet bubbled out to the surface in tremendous eruptions. (The same kind of thing happens to a thick soup if you leave it on the stove without stirring it.) The trapped gases, which included hydrogen, methane, nitrogen, ammonia, and carbon dioxide, formed the planet's early atmosphere. There was very little oxygen in that atmosphere.

HOT SPOT

Slowly, over millions of years, the atmosphere cooled and the first rains began to fall. Oceans formed. But Earth was still a very hot place. The surface temperature remained above 500°C for millions of years. It would be hard to imagine a more violent and lifeless environment. Volcanoes kept things cooking; lightning lit up the sky; and huge meteors struck the planet like bombs from outer space. But between 3.5 and 3.8 billion years ago, the deadly gases somehow combined to

give life a chance to develop. No one knows exactly how the first living things developed. Nor does anyone know how those first life-forms looked. But scientists do know that all forms of life have the same basic chemistry and the same genetic material. They also know that the first simple cells appeared in the form of bacteria about 3.5 billion years ago. Theories about the chemical origin of life are based on this knowledge.

A Big Ball of Soup

One explanation of how life may have started is the "primordial soup" theory. (The word primordial means first or earliest.) This theory is based on the fact that all life needs carbon-based proteins. According to this theory, the energy in lightning and sunlight caused gases in the atmosphere to react with one another and to form carbon-based compounds, which fell into the ocean that covered the planet. This mixture of compounds and ocean water was like an enormous soup whose ingredients were blending together. The compounds in the primordial soup gradually developed into living cells.

No one knows exactly how the first living things developed. Nor does anyone know how those first life-forms looked. But scientists do know that all forms of life have the same basic chemistry and the same genetic material.

To test this theory, scientists filled a chamber with the various gases that they think were in the early atmosphere. Then they ran electrical currents through this "atmosphere" to simulate the energy supplied by lightning. The electrical energy caused sugars and amino acids to form. Amino acids are the "building blocks" of proteins, and proteins and sugars are important materials for life. This experiment showed that such materials could be formed from simple chemicals and energy.

But the building blocks for life are not the same as life itself, just as a pile of bricks is not the same as a house. After all, proteins and sugars were also present in the ingredients that your teacher combined in the last demonstration, but those building blocks did not come to life. Something else must be needed. But what is it? The answer does not involve some missing ingredients. It involves some missing instructions.

Living things are organized. Every living thing comes from other living things that have the same structure and organization. The organizational instructions for building and maintaining every new cell in an organism are encoded in its genes. Genes are the hereditary material that a living thing receives from its parents. All genes are composed of **deoxyribonucleic acid (DNA)**. Every one of your cells follows instructions encoded in the structure of your DNA. A related compound called **ribonucleic acid (RNA)** helps your cells follow those instructions. Because DNA and RNA are necessary for the development of every new living thing, they were probably present at the origin of life on Earth.

Repeat That

How could the first DNA, RNA, and proteins have formed? One idea is that RNA formed first, followed by DNA and proteins. They may have developed in ponds with clay bottoms rich in iron. The surfaces of clay particles are composed of silicon, iron, and oxygen atoms arranged in a repeating geometrical pattern. DNA, RNA, and protein molecules are also composed of subunits arranged in a repeating pattern. These subunits are simple compounds that may have been present in the primordial soup. The molecules of these compounds have shapes and electrical charges that cause them to bind to clay particles. The

Just a Sideshow?

Could it be that life on earth began in hot, sulfurous waters at the bottom of the ocean? Many scientists believe that the first life-forms were archaebacteria that lived around hydrothermal vents on the seafloor, where the water pressure is so great that it would kill most living things.

AIT Video

Scientists once believed that all life was dependent on plants and bacteria that used the energy in sunlight to produce their own food. But archaebacteria get what they need to survive directly from the chemicals in their environment: hydrogen sulfide and methane.

The bacteria help to sustain newer life-forms, such as giant tube worms, which live near these vents. These enormous worms have no mouth or digestive system. A tube worm lives by absorbing nutrients through its tentacles and by getting energy from the bacteria living inside it. For their food, the bacteria use sulfuric compounds that the tube worm absorbs. In turn, the bacteria produce carbohydrates and proteins that the worm uses.

These deep-sea communities are still going strong after billions of years. Scientists think that the majority of the earth's living matter is in the oceans, most of it around hydrothermal vents. Looking at the number of species in each place, some people even wonder if life on land is merely an evolutionary sideshow to life in the sea.

clay can hold these small building-block molecules together, positioning them to form **polymers**, which are long chains of connected molecules, such as DNA and protein (see Figure 2). The DNA and RNA that directed the organization of the first living cells may have formed in this way.

Another theory is that the first DNA and RNA formed at **hydrothermal vents**. These vents are cracks in the ocean floor where heat from deep inside the planet is released under great pressure. The oldest known forms of life, the **archaebacteria**, still live around hydrothermal vents today. Unlike most forms of life, archaebacteria do not need energy from the sun or other organisms. They get the energy they need directly from the same kinds of volcanic sulfur and gases that bubbled out 3.8 billion years ago.

Millions of Chances

Although polymers could provide the structure that life needed to get started, it is not clear how these polymers first became capable of passing on their structure in DNA and RNA. But random combinations may have occurred for millions of years. With so much time for so many chance combinations, it is possible that some polymers appeared that could make copies of themselves. The most efficient of these self-copiers would have been the ones that could make the most copies of themselves from

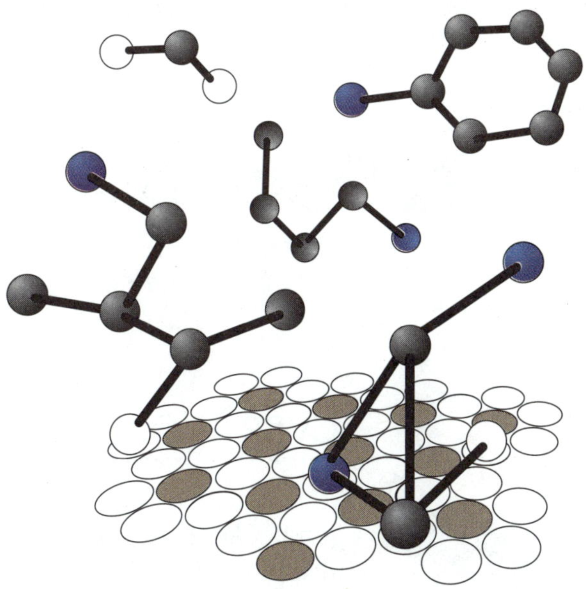

Figure 2: Small molecules binding to a clay particle
Note: The pattern of oxygen and silicon atoms on the surfaces of clay particles may have helped to organize the first DNA, RNA, and protein polymers.

the ingredients available in the primordial soup. Competition for chemical resources would have resulted in a process of natural selection that favored these polymers.

Life is also organized in other ways. For example, every living cell is surrounded by a selectively permeable membrane. This type of membrane controls which substances enter and leave the cell. Cells use this process to absorb and retain nutrients, expel wastes, and maintain an internal water content that is not too high or low. How did this process begin? No one can say for sure, but scientists know it is possible to create tiny droplets of fat or proteinlike substances that can absorb some materials but not others. These substances are not alive, but they show the possibility that primitive cells with similar properties could develop.

UNSOLVED MYSTERY

Despite all these theories, the basic question remains: How did the first bacteria come to life an estimated 3.5 billion years ago? Pressure from hydrothermal vents may have caused carbon compounds to organize into polymers, but how did these polymers form structures that could copy themselves? And how did mixtures of chemicals first become organized into cells? The mystery of how the first cells formed is not yet solved, but researchers are working on it.

ON YOUR OWN

Answer the following questions based on "The Origin of Life."

1. Why is it unlikely that the planet Earth could have supported life earlier than 3.8 billion years ago?
2. Sugar crystals "grow" in solutions of sugar: Loose molecules of sugar attach themselves to the crystal. How is this process different from making DNA and protein from hydrogen, nitrogen, oxygen, and carbon?
3. The temperatures at hydrothermal vents are extremely hot, but archaebacteria thrive there. How could you use this as evidence that life may have originated as early as four billion years ago?

INVESTIGATION 3

Ahhh, Oxygen

EVER WONDER?

You have seen how the earliest form of life could not have breathed oxygen because the earth had very little of it in the atmosphere. Today an abundance of life thrives on an atmosphere that is rich in oxygen. When did organisms begin to breathe oxygen? How did oxygen build up in the atmosphere? Did this buildup have anything to do with competition and natural selection? You will discover the answers to these questions in this investigation as you explore how important oxygen is to life on earth.

Materials List

- Beaker (250 mL)
- 2 flasks (250 mL)
- Yeast suspension in water
- 2 graduate cylinders (100 mL)
- Molasses-water mixture
- Laboratory pipet (1 mL)
- Pipet bulb
- Balloon
- Cotton ball
- Aquarium pump with air stone
- Masking tape for labels
- Pencil
- Stirring rod
- Safety goggles

Step-by-Step

Day 1

1. Put on your safety goggles.
2. Use your beaker to obtain 200 mL of molasses-water mixture.
3. Use a 100-mL graduate cylinder to transfer 100 mL of the molasses-water mixture to each of two 250-mL flasks.
4. Use a pipet to transfer 1 mL of yeast suspension to each flask. Swirl each flask gently to mix the contents after adding the yeast.
5. Label one flask "No Air." Stretch the balloon opening and place it over the top of this flask. Make sure you have a tight seal between the balloon and the flask.
6. Place the air stone, which is at the end of the aquarium-pump tube, on the bottom of the second flask. Make sure the flask is stable.
7. Plug the aquarium pump into an electrical outlet. Make sure it is bubbling air through the liquid inside the flask.
8. Plug the mouth of the flask loosely with a cotton ball. Do not push the cotton down into the flask. Label this flask "Air."
9. Set both flasks aside overnight in a warm place.
10. Clean up all the other equipment and put it away. Wash your hands well with soap and warm water before leaving the laboratory.

Day 2

1. Put on your safety goggles.
2. In your logbook, record any changes that you observe in the two flasks.

3. Examine the balloon, and record any changes you observe in it.
4. Unplug the aquarium pump.
5. Gently remove the cotton and the air stone from the "Air" flask. Try to remove as little of the contents as possible when you remove the air stone.
6. Gently remove the balloon from the "No Air" flask.
7. Swirl each flask with a gentle circular motion. Be careful not to spill the mixture. Indicate in your logbook whether each mixture is cloudy or clear.
8. Label one graduate cylinder "Air" and the other graduate cylinder "No Air."
9. Pour the contents of the "Air" flask into the "Air" graduate cylinder. Pour the contents of the "No Air" flask into the "No Air" graduate cylinder.
10. Set the graduate cylinders aside. Allow the cloudy material to settle to the bottom of each one.
11. Observe the cylinders as the mixtures settle. (The yeast cells will sink to the bottom in about 3–5 minutes.) In your logbook, record any differences you observe between the two mixtures.
12. Determine the volume of yeast in each cylinder by observing which measuring mark is closest to the top of the yeast. (The yeast is at the bottom of the cylinder.) In your logbook, record the volume of yeast in each cylinder.
13. Clean up your work station. Return all supplies to your teacher. Wash your hands well with soap and warm water before leaving the laboratory.

CAREER LINKS

Engineering Technologies

IBM

PETROLEUM GEOLOGIST

Petroleum geologists search for oil and gas deposits under the earth's surface. They use their knowledge of geology, geophysics, seismology, and computer-imaging techniques to map the areas that they explore. They must also consider conservation and environmental issues. Petroleum geologists design and improve drilling methods for extracting petroleum from the earth. Some also work in the extraction process itself or in the production of petroleum products. A petroleum geologist needs at least a bachelor's degree in the geological sciences. For more information, write:

The Geological Society of America
3300 Penrose Place
Boulder, CO 80301
http://www.geosociety.org

Related occupations: geological engineer, paleontologist, hydrologist, geochemist, civil engineer

Talk It Over

Work with your partner to answer the following questions.

1 Why was it necessary to use a water-molasses mixture with the yeast? (Hint: Remember that yeast are living organisms.)

2 What was the most important difference between the environments that the yeast cells experienced in the two flasks? Based on your knowledge of living things, give the reasons that you think this difference could be important.

3 Did you observe any change in the balloon? If so, explain what you think caused this change. (Hint: Think about your response to Question 2.)

4 Did the flasks look any different on the second day when you swirled them around? If so, describe the difference and tell what you think caused it.

5 Did more yeast grow in one flask than in the other (as shown by the amount at the bottom of each graduate cylinder)? What do these results tell you about the growth of yeast?

6 Suppose that early life on earth contained two colonies of organisms like the yeast in your investigation. One colony lived in an environment rich in oxygen ("air"), and the other colony lived in a low-oxygen environment. Which organisms would have a better reproduction rate? How would this affect their chances of long-term survival?

7 Would the more successful colony of organisms, as described in Question 6, be likely to threaten the survival of the other colony? Explain your thinking.

Spread the Word

Work with your partner to present your findings to the rest of the class. You could start by describing your investigation. Explain the purpose of each component that you used: the yeast, the molasses-water mixture, the aquarium pump, the cotton ball, and the balloon. Explain how your results lead you to conclude something about the role of oxygen in yeast growth. Highlight the key difference between the two flasks, and explain how this difference is important to your conclusions. Suggest ways that this could play a role in the development of early life on earth.

List several careers in which workers need to understand what you discovered in this investigation.

READ ALL ABOUT IT!

The First Forms of Life

NOT all organisms need oxygen to survive. You discovered this during Investigation 3. Organisms that can obtain and use energy without needing oxygen are called **anaerobes**, a term meaning "life without air." Because the early atmosphere of the earth lacked oxygen, the first forms of life to develop on earth were anaerobes.

Fossils from western Australia and southern Africa show that the first anaerobes were primitive bacteria. Bacteria are **prokaryotes**, simple cells without a nucleus or other specialized parts. Their DNA is loosely gathered into a ring within the cell. Many similar kinds of prokaryotes exist today. They are very small, usually made of a single cell that is only about one-tenth the size of plant and animal cells.

The available physical evidence indicates that life has existed on earth for about 3.5 billion years. For most of this time (about two billion years), prokaryotes ruled the planet. This does not mean there was no diversity of life two billion years ago. Early bacteria had many varieties, just as they do now. But those early types of bacteria obtained their energy in a different way than today's bacteria do.

Researchers believe that the earliest bacteria received their energy directly from elements in the environment. They absorbed such molecules as carbon dioxide and hydrogen sulfide. Bacteria that can process chemicals in this way do not need oxygen. In fact, oxygen is poisonous to most of them. Such bacteria can thrive in hot, acidic environments that would kill most other organisms.

GETTING A DATE

If you want to know when a coin was minted, you can look at the date on the front. But how can you tell how many years ago a fossilized animal lived? Archaeologists had trouble answering this until the 1950s, when Walter F. Libby developed carbon-14 dating—a technique that earned him a Nobel Prize.

Carbon-14 is a radioactive form of carbon that is present in a small part of the carbon dioxide in the atmosphere. Plant life absorbs the carbon dioxide. Plants are eaten by animals, who in turn are consumed by other animals. As a result, the fossils of all these living things contain carbon-14 in the same percentage that it was present in the atmosphere at the time they were alive.

Scientists can use carbon dating to determine the age of a fossil because they know how long it takes the radioactivity in carbon-14 to decay. The radioactive half-life of carbon-14 is 5,730 years. This means that half of its original radioactivity took 5,730 years to disappear. Then it took another 5,730 years for half of the remaining radioactivity to disappear—and so on. It will always take 5,730 years for half of any remainder to disappear.

Carbon-14's rate of decay makes it useful for dating specimens up to 70,000 years old. For older specimens, a form of potassium (with a half-life of 1.3 billion years) or a form of uranium (half-life of 4.5 billion years) is used. By measuring the amount of radioactivity in a fossil, scientists can determine how long ago the organism lived.

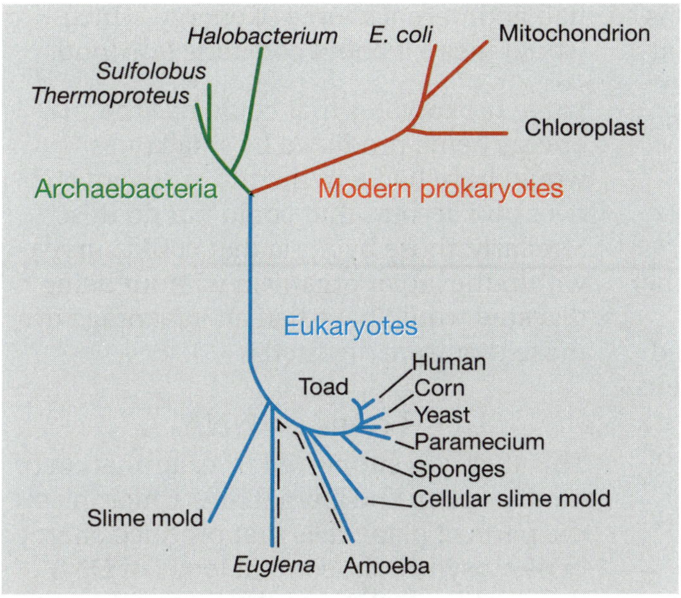

Distances between species on this diagram show how closely they are related. (The broken line is an example.) Animals, plants, and fungi (yeast) are more closely related to one another than many kinds of bacteria are.

For millions of years these bacteria grew in numbers and consumed the chemicals in their environment. This meant that they were competing for a shrinking supply of chemical resources. The prokaryotes that used the more plentiful molecules—or used resources most efficiently—had an advantage over those that did not. During this long-term competition to obtain energy, the process of natural selection favored variations that enabled bacteria to obtain energy in a different way.

Birth of the Blues

One type of prokaryote, **cyanobacteria** (bluish-green bacteria), developed a way to harness energy from sunlight. Cyanobacteria use light energy to convert carbon dioxide (CO_2) from the environment into carbon compounds that are necessary for life—compounds such as proteins and carbohydrates. A waste product of this process is oxygen (O_2). You may recognize this process as **photosynthesis**, which takes place in most of today's plants. As the first life-form to use this process, cyanobacteria can be considered the ancestors of all plants.

The success of cyanobacteria changed the world. With sunlight as a plentiful source of energy, cyanobacteria thrived and spread to many different environments. As they increased in numbers, they produced more and more oxygen. Chemical analysis of rocks shows that by 2.5 billion years ago, oxygen had accumulated in the atmosphere.

Oxygen Lovers

With an increasing supply of oxygen, organisms that developed ways to use oxygen had access to a new and growing resource. Variations began to appear, enabling some organisms to use oxygen to produce energy. These organisms are called **aerobes**. The oxygen-loving aerobes did not compete with the anaerobes for chemicals; therefore, the two types of bacteria thrived side by side. But as the number of aerobes grew and the amount of oxygen in the atmosphere increased, many of the anaerobic prokaryotes died out. Anaerobes that survived were probably protected by mud, deep water, or other environments that contained little oxygen, or they may have developed a tolerance for oxygen. Some organisms, such as many yeasts and molds, can survive in both high- and low-oxygen environments.

The success of cyanobacteria changed the world.

As you discovered during Investigation 3, it is more efficient to produce energy with oxygen than without it. The yeast that did not have a supply of air, which contains oxygen and other gases, did not grow as much as the yeast that did have access to air. The yeast cells that used oxygen to digest the sugar in the molasses released about 18 times as much food energy as the yeast cells that did not use oxygen. In other words, aerobic digestion is far more efficient than anaerobic digestion. Because aerobes are more efficient, they need less food to survive. It is easy to see how natural selection favored the development of aerobes.

The rise of aerobes was a major change in the history of life, but an even greater change occurred during the next billion years. It is not clear when or how it happened, but at some point about 1.5 billion years ago an entirely new life-form developed. Known as **eukaryotes**, which means "true nucleus," these organisms have a different kind of cell than prokaryotes have (see Figure 3).

Eukaryotic cells contain a nucleus and other parts that are specialized to do various tasks. The DNA of eukaryotes is contained within the membrane of every cell's nucleus. The appearance of eukaryotes was a tremendous development in the history of life because all organisms alive today—except for bacteria—have eukaryotic cells.

Well Organized

When a cell is organized by many parts or "little organs," each one doing its own task, the cell is more efficient at using energy and reproducing itself. This fact explains why eukaryotes became successful. But how did they first form? One theory is called the **endosymbiont hypothesis** (see Figure 4). This theory suggests that the specialized parts of eukaryotes arose when different kinds of prokaryotes combined. (The prefix "endo" means inside, and a "symbiont" is an organism that lives in close relation with another organism.) For example, a large prokaryote that obtained energy by swallowing up small ones may have absorbed several kinds of photosynthetic bacteria. Some of these may have survived and continued producing food with energy from light. The large prokaryote would then have had an internal source of energy, which would mean it could consume less food.

Large prokaryotes that could use the energy being produced by small ones would have had a tremendous advantage over prokaryotes that could not do this. Similarly, those bacteria that could survive within the larger organism without being digested would have had an advantage over those that were digested.

Loops of DNA

This theory is supported by examination of modern plant and animal cells. Chloroplasts, the parts of plant cells that produce energy in photosynthesis, contain loops of DNA that are like the DNA in the first prokaryotes that used sunlight to produce energy. In addition, all eukaryotic cells with looped DNA have other specialized parts that researchers think may be like the first aerobes. If large prokaryotes developed the ability to live with

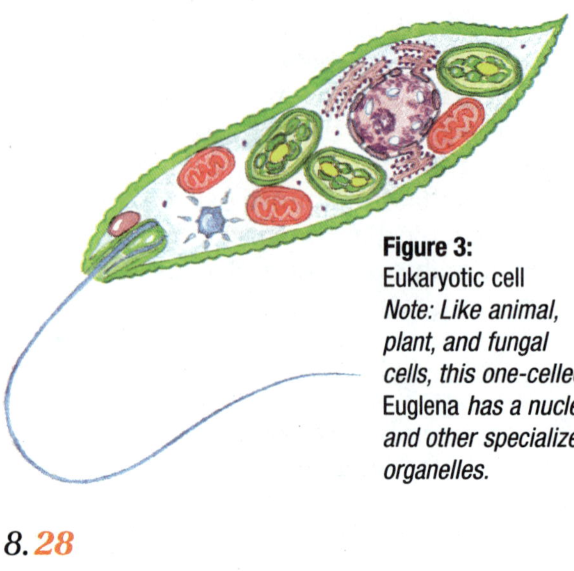

Figure 3: Eukaryotic cell
Note: Like animal, plant, and fungal cells, this one-celled Euglena *has a nucleus and other specialized organelles.*

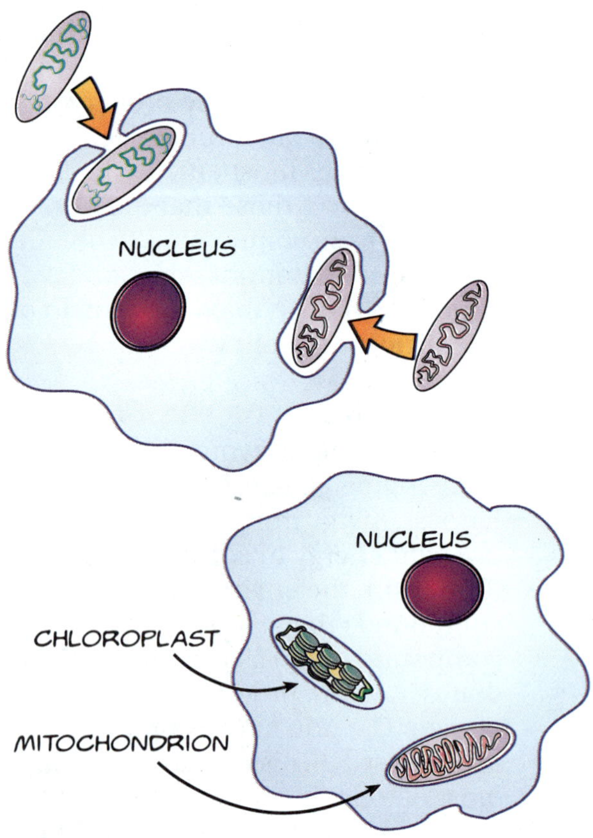

Figure 4: The endosymbiont hypothesis
Note: The first eukaryotes may have formed when primitive cells with nuclei swallowed up other cells but did not digest them.

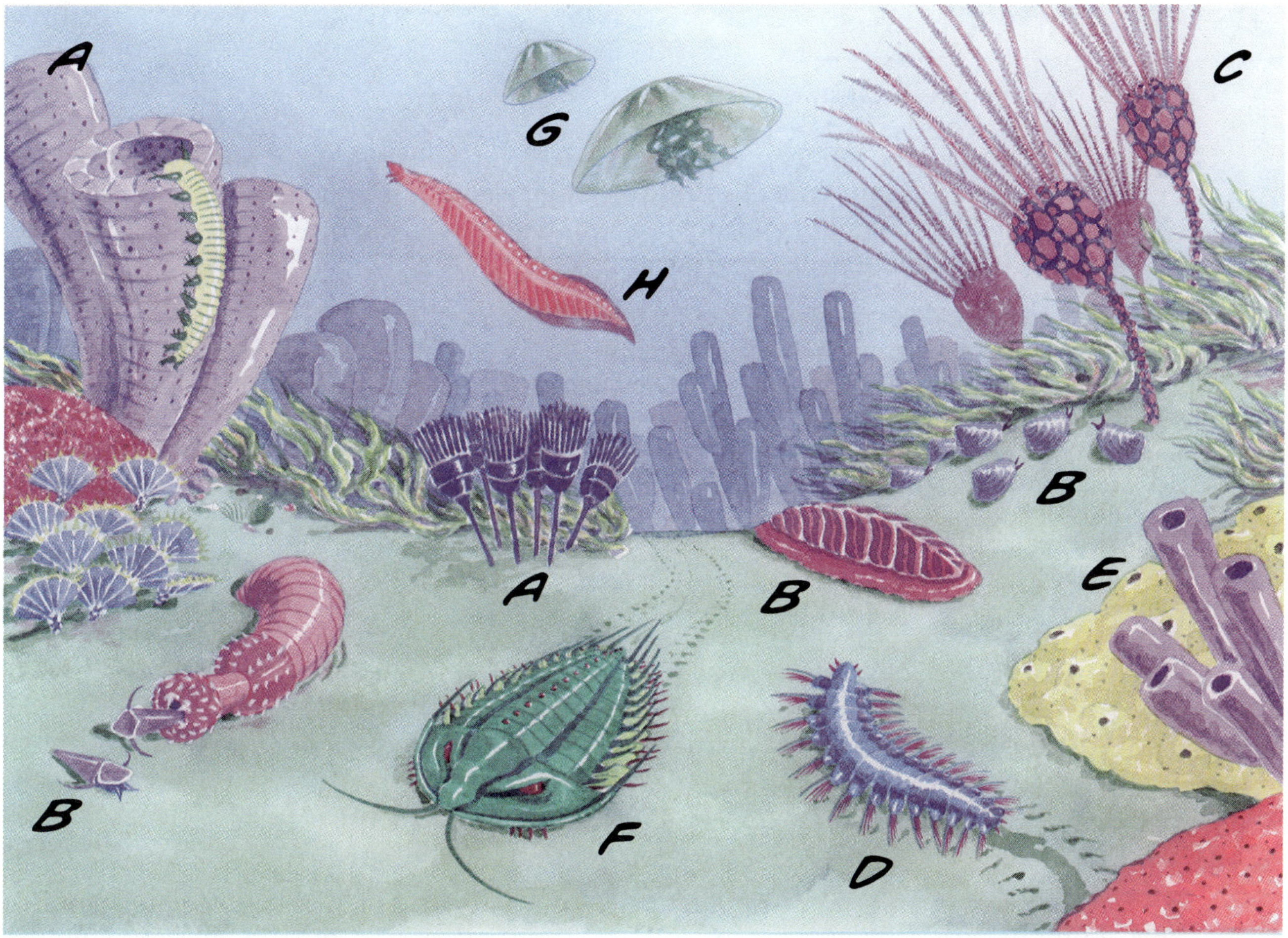

Cambrian animals were extremely diverse. They include both extinct groups (A) and relatives of today's clams and snails (B), sea stars (C), earthworms (D), sponges (E), shrimp (F), jellyfish (G), and possibly the ancestors of the vertebrates (H).

different kinds of smaller prokaryotes inside them, this would explain the origin of the eukaryotic cells of today's animals, plants, fungi, and many one-celled organisms.

Some researchers reject the endosymbiont hypothesis. They point out that the "little organs" in today's cells cannot live by themselves, as prokaryotes can. Therefore, they argue, eukaryotes must have developed in some other way. Perhaps over hundreds of millions of years of variations, some prokaryotes developed a more complicated cellular structure. Because structures that separated parts of the cell would be beneficial to the organism, these structures would be passed on from generation to generation. Eventually, after eons of variation and selection, cells that are similar to those of modern plants and animals may have developed.

Soon the oceans were swarming with many kinds of algae and fish, and the door was open for life on land.

However it happened, the appearance of eukaryotes led to an explosion in the variety of life. Variations are vital to the success of a species. Because eukaryotes have many more structures than prokaryotes, many more parts can vary over generations. Examination of fossils shows that in less than a billion years, these cells gave rise to the first ocean animals that relied on plants or other creatures for food. Soon the oceans were swarming with many kinds of algae and fish, and the door was open for life on land.

On Your Own

Answer the following questions based on "The First Forms of Life."

1 If you look at all the life-forms that exist today, you will see more prokaryotes than all eukaryotes combined—all algae, fungi, plants, and animals. Prokaryotes are also much more different from one another than eukaryotes are. How would you explain this?

2 In many businesses, people are trained to do specific jobs. The various departments in a company are groups of people doing similar tasks. The company as a whole depends on each department's performing certain kinds of jobs. How is a company like a eukaryote?

3 Organisms have been discovered living inside volcanoes, where it is very hot and there is little oxygen. What kinds of organisms do you think these are? Give your reasons.

Section Wrap-up

When Bacteria Ruled the World

For centuries people have wondered about the possibility that life exists on other planets. Unmanned space probes have traveled to Venus, Mars, Jupiter, and the outer planets. These probes have provided vivid pictures of the harsh conditions that exist on other planets and moons in the solar system. It is hard to believe that life could exist in atmospheres that have little oxygen or on surfaces that are racked with volcanic eruptions and hurricanes. But similar conditions probably existed on the planet Earth billions of years ago, and life was able to develop here.

Hot, Violent

Earth was once terribly hot and violent. Surface temperatures rose above 500°C as the planet released the heat that had been generated as it formed. For hundreds of millions of years, lightning storms lit the skies and volcanoes spewed dust and gases into the air. Then, when the world finally cooled down a bit, rains pounded the surface for several more million years. Even as the oceans were forming, Earth's surface continued to be split apart by earthquakes as the planet went on venting its heat.

Scientists have tried to simulate these conditions in the laboratory. The results have been amazing. When they sent electrical current through gases such as methane, water vapor, and hydrogen sulfide, they discovered that amino acids, the building blocks of proteins, were formed. Proteins are essential for life. These experiments showed that lightning could have acted on gases in the atmosphere to produce the same kinds of substances. Even in the harsh conditions that existed when the world was young—when hardly any oxygen was in the atmosphere—the ingredients for life may have been produced in a similar way.

Getting It Together

How did these ingredients come together as life? This is a great mystery that scientists are still trying to solve. Researchers have experimented with these ingredients in many different situations. They have made some progress. For example, they have shown that polymers will form randomly from the ingredients under certain circumstances. Polymers are necessary for life. Further tests have shown that some of these polymers can make copies of themselves. This self-copying ability—the ability to pass on the structure of life—is essential for life to continue once it has formed. The structure of every organism is encoded in its DNA and RNA. Though interesting theories about the origin of proteins, DNA, and RNA exist, no one is sure how these substances first formed.

Get This . . .

Bacteria are **prokaryotes**, simple cells without a nucleus or other specialized parts. Fossil evidence shows that primitive bacteria first appeared about 3.5 billion years ago. All organisms alive today—except bacteria—have eukaryotic cells. **Eukaryotes** have cells with a nucleus and other parts that are specialized to do various tasks.

How did the first life-forms look? Ideas about the first kinds of life are based on today's organisms. The oldest and most primitive forms of life exist near hydrothermal vents on the ocean floor. The bacteria that live there demonstrate that life can exist under extremely harsh conditions. Hydrothermal vents produce great heat, and the water pressure at these depths is tremendous. Yet life goes on, despite the absence of oxygen and sunlight.

These primitive bacteria are prokaryotes, which belong to the largest kingdom of life. Their simple cells lack the nucleus and other specialized structures of plants and animals. The DNA of prokaryotes is loosely gathered into a ring within the cell. Prokaryotes first appeared about 3.5 billion years ago, and for the next two billion years these simple organisms ruled the world. Many different types of prokaryotes developed, all competing for energy and resources in the ancient oceans.

Lighten Up

During that two-billion-year span, an important new life-form, cyanobacteria, developed among the prokaryotes. Cyanobacteria had an advantage over other prokaryotes because they could use energy from sunlight. These first ancestors of plants made food through photosynthesis, releasing oxygen into the atmosphere. Their ability to make oxygen gave them another advantage: As the level of oxygen increased, the prokaryotes that could not tolerate oxygen suffered. These anaerobes either died or developed variations that enabled them to live in the changing environment.

Earth was filled with many kinds of bacteria. As the air and oceans cooled, some prokaryotes could survive only deep within the crust or inside volcanoes, where conditions were similar to those that existed when the planet was young.

The process of adaptive variation did produce more kinds of organisms. Aerobes developed. These bacteria use oxygen to process food. Using oxygen to make food provides much more energy than making food without it. Aerobes are more efficient than anaerobes. Because they needed much less food to survive, the environment could support many more of them. Although oxygen-loving aerobes did not compete directly with oxygen-hating anaerobes, natural selection favored the aerobes, and aerobic species began to outnumber the anaerobic ones.

That's Cool

Earth was filled with many kinds of bacteria. As the air and oceans cooled, some prokaryotes could survive only deep within the crust or inside volcanoes, where conditions were similar to those that existed when the planet was young. Cyanobacteria spread across the surface, thriving where they could get a lot of sunlight. Other bacteria developed ways to absorb one another, getting energy from the organisms they consumed. These prokaryotes were the first predators. Many researchers believe that these simple, microscopic predators gave rise to a whole new kind of life, which in turn led to the development of all plants and animals. When larger prokaryotes consumed smaller ones, some of the smaller ones may have survived and continued making food with light or oxygen. Organisms that could survive on food produced by other prokaryotes inside them would be able to survive even when prey was hard to find.

The Origin of Eukaryotes

It is possible that these endosymbionts developed into the first complex cells with specialized parts: eukaryotes. This is only one theory that seeks to explain the origin of eukaryotes, whose fossils date back 1.5 billion years.

Once cells developed different parts that performed specific functions, variations could arise in many different ways. Soon there were hundreds of species of eukaryotes, filling the oceans with sponges, algae, corals, and starfish.

This explosion of new life-forms continued. The number of species rapidly increased, making it possible for even more new species to develop. As the process of natural selection acted on this ever-increasing number of early plants and animals, the stage was set for the development of more complex forms of life, including the plants and animals that exist today.

Get This . . .

- Organisms that can obtain and use energy without needing oxygen are known as **anaerobes**—a term meaning "life without air."
- Organisms that use oxygen to produce energy are called **aerobes**.

On Your Own

Respond to the following items based on "When Bacteria Ruled the World."

1. If you wanted to find out whether life could develop on other planets or their moons, what could you do?

2. Suppose scientists produced DNA in the laboratory in conditions similar to those that existed 3.5 billion years ago. Would this be good evidence for how life began? Explain.

3. The oldest fossils indicate that life originated about 3.5 billion years ago (about a billion years after the plant Earth formed). Would it be possible to find a fossil that was more than two billion years older than the oldest known fossils? In other words, could a 5.5-billion-year-old fossil be found? (Hint: Recall what you know about the early conditions on Earth.) Would it be likely that someone would find such a fossil? Explain your answer.

4. Although most species are eukaryotic, most of the world's living matter consists of anaerobic bacteria. Explain why anaerobic bacteria did not die out after the rise of the more complex eukaryotes.

Career Links

Health & Human Services

World Bank

Epidemiologist

Epidemiologists are like medical detectives. They figure out what causes and transmits diseases, especially ones that are epidemic. Epidemics are outbreaks of disease that spread rapidly. Epidemiologists study contagious diseases, the health risks of harmful environments, and poor health practices. They also try to learn how disease-causing organisms become resistant to certain drugs, such as antibiotics. Public education is an important part of the work of these health specialists. For example, they publicize the need for infant and childhood vaccinations; they also educate people about how to avoid HIV infection and other contagious diseases. If you want to be an epidemiologist, you will need an advanced degree. For more information, write:

American Public Health Association
1015 Fifteenth Street, N.W.
Washington, DC 20005-2605
http://www.apha.org

Related occupations: biostatistician, public health official, geneticist, microbiologist, parasitologist

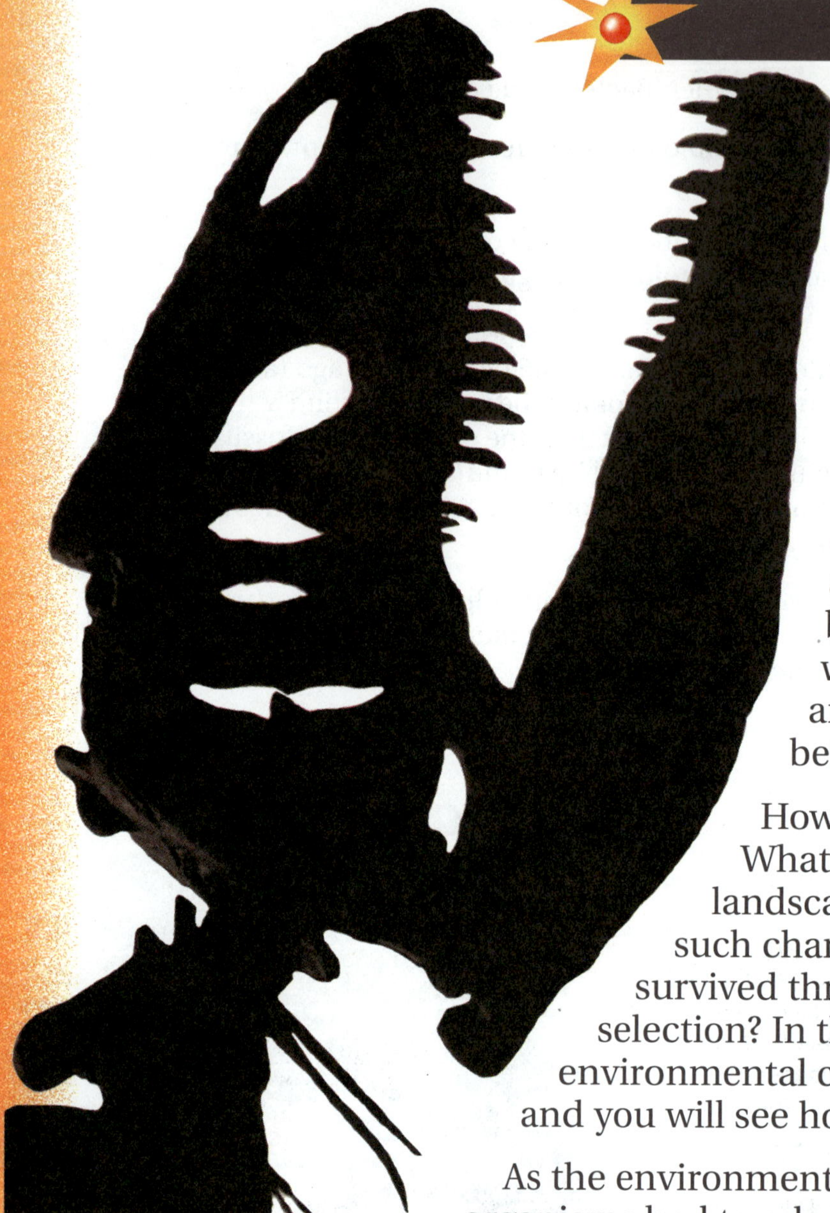

Section 3

THE HISTORY OF LIFE

HAVE you ever thought about what kinds of things lived on Earth millions of years ago? You have probably seen pictures of dinosaurs. Some types of dinosaurs grew larger than any animals living today, except the biggest whales. Can you imagine what other kinds of species lived and died in the distant past, long before you were born?

How did today's species develop? What caused others to die out? Did the landscape or the climate change? Could such changes determine which variations survived through the process of natural selection? In this section you will examine the environmental changes that led to new species, and you will see how they spread around the world.

As the environment changed over millions of years, organisms had to adapt—they had to live in new and different ways. What strategies did various species use to survive? How did their changes produce the species of today? By the end of this section, you will understand how the species of the distant past led to those you can see today.

FREEZE FRAME

Now that you have seen the video segment, discuss the following on-screen questions with the rest of the class.

1. How did many ancient species die out?
2. Do most kinds of organisms live only in certain places today? How can this be explained?

On the Beach

Ever Wonder?

The first forms of life on earth probably developed in the oceans. So then how did life get started on dry land? After all, a fish can't live out of water! What does a plant or an animal need to survive and reproduce on land? The organisms that live on land are quite different from one another. Do they all have anything in common? In this investigation you will examine how living things have adapted to life out of water.

Materials List

- Celery stalk, including leaves
- Knife
- Food coloring added to water in jars
- Chicken egg
- Bowl
- Fish eggs
- Paper towels
- Clean glass or microscope slide
- Dried insects
- Apple or cucumber
- Beaker of water
- Dropper pipet (medicine dropper)
- Houseplant in pot
- Water plant in water

Step-by-Step

In this investigation you will observe samples of some organisms that live in water and some that live on land. As you examine each group of samples, think about the characteristics that help each plant or animal live on land or in water. In your logbook, record your thoughts and observations about each of these characteristics.

1. Cut a 1-cm piece of celery off the bottom of the stalk. *Caution: Be very careful not to cut yourself or your partner(s) with the knife.*

2. Place the celery stalk, with its cut end down, into a jar of colored water. Allow the stalk to stand in the water while you visit the lab stations around the room.

3. Visit the three lab stations in any order. Follow the instructions provided for each station.

Station 1: Eggs

- Compare a chicken egg with fish eggs. Compare the appearance, size, surface, and other characteristics of the two types of eggs. Notice what is the same or different about them.
- In your logbook, record your observations about the eggs. Also record your thoughts on how the eggs help the species that produce them to survive on land or in water.
- Leave the chicken egg and some of the fish eggs exposed on a dry paper towel at this station.

Station 2: Skins

- Smear your fingers across the clean glass or microscope slide.
- Examine the glass or slide carefully. If you see nothing on it, rub your finger on it again. What do you see?
- Write your observations in your logbook. Also write down your ideas on how what you have observed may be related to the ability of people to survive out of water.
- Examine the insect bodies and the apple or cucumber.
- Gently scrape the surface of the apple or cucumber with the knife. *Caution: Be careful not to cut yourself.* Observe what comes off onto the blade. Think about what keeps these living things from drying out. Record your observations and ideas in your logbook.
- Use the dropper pipet to put a drop of water on the apple or cucumber. Record your observations.

Station 3: Plants

- Examine the water plant. Think about where this organism lives and what supports it (holds it up). Also think about how the plant takes in nutrients and water. Record your observations and ideas in your logbook.
- Examine the houseplant—look at all parts of the plant. Gently remove it from the pot. Think about what the roots might do for the plant. Also think about what keeps the plant from drying out. Record your observations and ideas in your logbook.

4 Return to your work area. Look at the celery stalk, especially the base. Observe what has happened to the colored water. Record your observations.

5 Examine the eggs at Station 1 again. In your logbook, record any changes that you observe in the eggs.

6 Gently break a chicken egg in a bowl. Be careful not to break the yolk.

7 Look carefully at the shell. Think about how the membrane on the inside of the shell might help a chick survive inside the egg. Consider whether the inside of the chicken egg looks like a fish egg. Record your observations and ideas about these two kinds of eggs.

8 Clean up your work area, and return all your materials. *Caution: Wash your hands well with hot water and soap before leaving the room.*

Talk It Over

Work with your partner(s) to answer the following questions.

1. Compare the outsides and the insides of the chicken and fish eggs. How does the chicken egg help the chick inside to survive on land? How could the membrane inside the shell of the chicken egg help a chick to survive?

2. Compare the surfaces of plants and animals that live in water with those that live on land. In what ways are these surfaces different? How could these differences help plants and animals to survive in the different habitats?

A Living Fossil

Humans get their oxygen by breathing air. Fish use their gills to breathe—right?

Not always. The lungfish has either one or two lungs and can breathe through its mouth.

The lungfish is a large carnivorous animal that usually lives in water. It is born with gills. In some species, the gills are reabsorbed back into its body after it becomes an adult. But during periods without rain, it relies on its ability to breathe. It can burrow into soft mud and live there for years at a time, if necessary.

This uniquely adapted animal is called a living fossil because lungfish species reached their peak about 400 million years ago. Today only a handful of species survive in Australia, Africa, and South America. Scientists fear that the lungfish is headed for extinction.

3. What problems could an organism that lives in water have if it were moved onto the land?

4. What problems could an organism that lives on land have if it were moved into the water?

5. Many land plants produce seeds with hard, dry coats, but seaweed and other water plants seldom do so. What are the reasons for this?

Spread the Word

Work with your partner(s) to share what you have learned with the other students in your class.

Make a list of various functions that organisms must perform in order to survive and carry on their species. Describe how each organism that you observed has adapted to carry out those functions. Then list the characteristics that the water organisms have in common with the land organisms, and explain how these common traits differ. Be sure to tell how you think early species had to adapt to live on the land. To do this, think about what is needed to live on land.

Describe several types of work in which people need to understand the different characteristics of land and water organisms.

Get Out of the Water

AFTER taking a long bath or washing dishes, did you ever notice how dry your skin felt? Your fingertips were rough and wrinkled. How could water make anything drier?

Most species that live on land, including humans, have a thin layer of oil or wax on their skin. This helps keep them from drying out. You observed this during Investigation 4 when you smeared some of your skin oil onto a piece of glass. If you spend a long time in the water, you wash this oil off your skin, which then begins to dry out.

HIGH AND DRY

When early organisms spread from sea to land, the first problem they faced was how to protect themselves against dehydration (drying out). Water is essential to life, and all

land organisms must be able to hold water inside them. In comparing water plants with house plants, you saw that water plants do not have thick stems, highly branched roots or veins, and firm waxy leaves—traits that allow land plants to retain water. If a water plant were washed ashore, it would quickly dry out and die. To survive on land, water plants had to develop new traits. Over time, they developed roots and veins that carry water and dissolved nutrients from the soil up through the rest of their bodies. You saw an example of this adaptation to land when you observed the transport of colored water up a celery stalk. This and other adaptations enable some plants, such as cactuses, to live in extremely dry places.

Frogs, salamanders, and other amphibians have characteristics of both fish and reptiles.

How did animals overcome the problem of dehydration? Some species have not overcome it completely. Frogs, salamanders, and other **amphibians** have characteristics of both fish and reptiles. While they can live on land, they must have wet environments. Frogs lay their eggs in water, and after the eggs hatch, the larvas (tadpoles) continue living in water, breathing through external gills, as fish do. Later, when tadpoles develop into frogs, the gills become internal and the frogs can live on land. Many types of worms also need a moist environment.

WATERPROOFING

Unlike frogs and worms, most land dwellers developed a waterproof layer at the surface of their bodies. Plants developed a waxy layer on their leaves and stems. Reptiles, birds, and mammals developed oil on their feathers and skins. Arthropods, such as the insects you examined, have an **exoskeleton**, a wax-coated layer of body armor on the outside.

A waterproof barrier at the surface of the body raises another problem for living on land. Living things need to exchange gases (oxygen and carbon dioxide) with their environments. Many aquatic species absorb and release gases in water through the entire surface of the body. But animals that live in air must have a moist surface in which these gases can dissolve before they can enter the body. Frogs, earthworms, and some other animals that live in moist environments use their entire skin for gas exchange in this way.

Exchanging Gases

Other land species use internal moist surfaces to exchange gases. The leaves of plants have tiny pores that allow gases to move in and out through the moist surfaces of the cells within the leaves (see Figure 5). Reptiles, birds, and mammals have moist internal lungs that take in oxygen and expel carbon dioxide (see Figure 6). Insects have air tubes (tracheas) that carry air directly to all cells of the body (see Figure 7).

All these systems involve the controlled flow or diffusion of air through a small opening. Each system provides a way for an organism

Self-Sealing Toads

If you were a toad, where would you live? You'd probably enjoy living by a nice pond, wouldn't you?

Toads, being amphibians, need to keep their skin moist. Because of this, they usually live near water. They also lay their eggs in ponds. However, one type of toad manages to live in the desert.

The desert toad has adapted to its dry environment in some interesting ways. Like many desert animals, this toad copes with extreme temperatures by digging itself a burrow. To preserve its vital moisture, the toad coats itself with mucus and becomes dormant. during extended dry periods. The mucus protects its skin, locking in moisture until rain comes.

Figure 5: Gas exchange in leaves
Note: Tiny openings (stomata) for carbon dioxide and oxygen restrict evaporation of water from leaves.

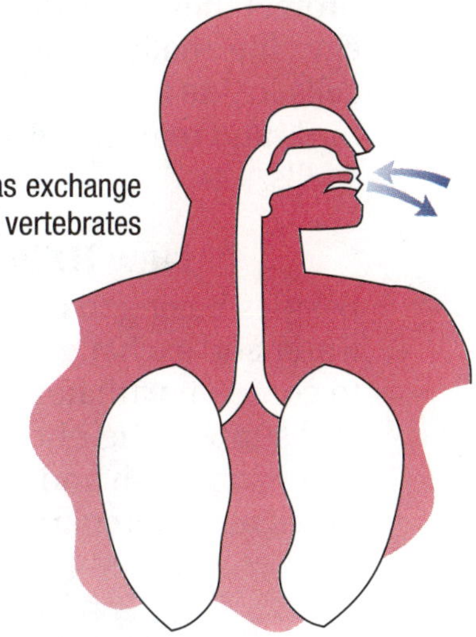

Figure 6: Gas exchange in land vertebrates

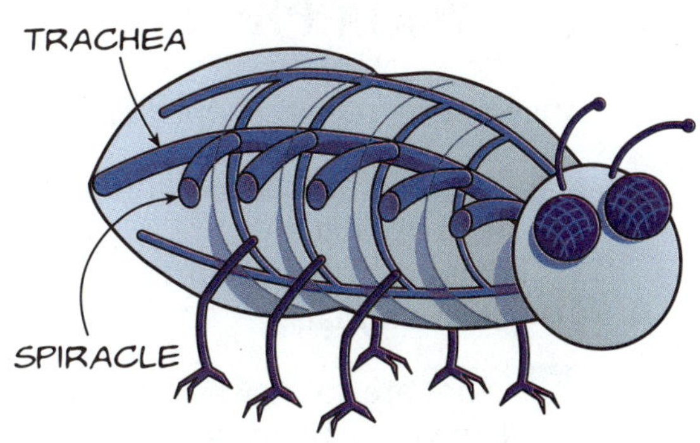

Figure 7: Gas exchange in insects
Note: Air passes in and out of the tracheas through tiny openings (spiracles).

GET THIS . . .

To protect themselves against dehydration, most species that live on land, including humans, have a thin layer of oil or wax on their skin.

to exchange gases with the environment but still keep water inside its body.

The need to retain water is also important for reproduction. Think about what happened to the eggs in the investigation. When removed from water, fish and amphibian eggs dry out and die. The same problem faces plants such as mosses, which must rely on water to carry their reproductive cells. Amphibians must return to water to reproduce. Successful land dwellers are those whose eggs can retain water and nutrients.

To protect reproductive cells and developing offspring, land plants developed seeds and pollen with waterproof coverings. Mammals protect their reproductive cells by fertilizing eggs internally: Males deposit their sperm cells inside females. Other animals, such as birds and reptiles, have eggs with waterproof shells.

NEW REQUIREMENTS

Besides having to develop ways to retain water and exchange gases, land animals had to come up with new ways to move around and support their bodies. They needed legs and wings. Fish and many other organisms that live in water are supported mainly by the water around them. Organisms that moved onto land needed new support systems. Most land plants developed roots that hold them firmly in the ground and woody stems made of cells with rigid walls that support them above ground. The insect's exoskeleton not only keeps it from drying out but also does an excellent job of supporting its body. Reptiles, birds, and mammals have a complex system of bones and muscles that enables them to run, jump, and fly.

Surrounded by water, where the temperature is nearly constant, animals do not need to regulate their internal temperatures. But on land, where it may be cold at night and warm during the day, animals had to adapt. Some animals, such as amphibians and reptiles, are cold-blooded. As a result, they must depend on the outside environment to regulate their temperature. If the weather is too hot, a reptile hides in the shade. If the weather gets too cold, the reptile warms itself in the sun.

Most of these animals can survive only in warmer parts of the world. Tropical rain forests are filled with many species of lizards and snakes, but few of these organisms live high in the mountains. On the other hand, birds and mammals can live anywhere because they are warm-blooded. They have an internal heat source that keeps them at a uniform temperature. Because of this, you can find ducks and dogs from the equator to Alaska.

ON YOUR OWN

Answer the following questions based on "Get Out of the Water."

1. Humans and other land animals with backbones develop "gill slits" in their necks while they are in the womb. These gill slits appear early in life but disappear before birth. How does this fact provide evidence of the origin of such animals?

2. What are some ways that land plants spread their seeds to other places? How can you explain the fact that water plants do not have these adaptations?

3. What problem has living on land created for the nutrition of developing embryos? What are some ways that animals have solved this problem?

The End

EVER WONDER?

Do you like to follow your favorite team through the championship playoffs at the end of the season? Maybe there's a tennis player, gymnast, or ice-skater that you always watch when a tournament is on TV. You can usually predict how well most of the competitors will do, but sometimes you get a surprise: The team that was expected to win gets knocked out of the playoffs, or an injury takes a great player out of the tourney.

Something similar happens in nature. Millions of different kinds of organisms have lived on earth, but many species have not survived. It is not always clear why one species survives longer than another. In this investigation you will play a game to explore what types of factors contribute to the extinction of a species.

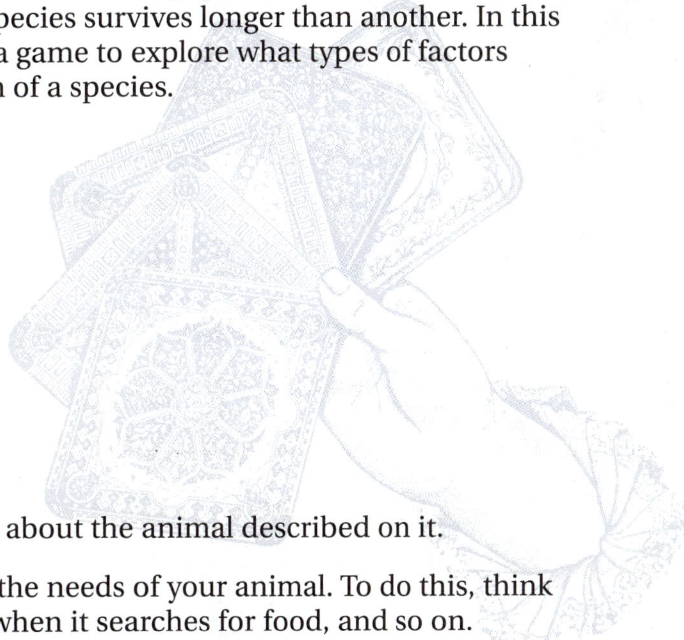

MATERIALS LIST

- Fact cards
- 6 coins

STEP-BY-STEP

1. Examine your fact card, and think about the animal described on it.
2. In your logbook, make a list of all the needs of your animal. To do this, think about what it eats, where it lives, when it searches for food, and so on.
3. Your teacher will announce an environmental change. Check your fact card to see if your animal is affected by this change.
4. If the environmental change—for example, reduced food—would be harmful to your animal, flip your coin to see if your animal survives. If the coin lands "heads," your animal dies; if it lands "tails," your animal survives. Record the result in your logbook. (If your animal dies, continue to record what happens to the others.)
5. Check with the other members of your group to see which animals have lived and which have died. Record this information.
6. Your animal's life may depend on the life of one or more of the other animals in your group. If your animal is still alive but all the other animals on which it depends die off, then flip your coin to see if your animal survives by migrating to another group. If the coin is "heads," your animal does not survive; if the

coin is "tails," your animal does survive by moving to another group—but only if you can find a group that still has the resources your animal needs. If you cannot find such a group, then your animal dies too. Record what happens.

7. Your teacher will announce more environmental changes. After each change, repeat the procedure you followed in Steps 3–6.

8. After all environmental changes have occurred, report to your teacher the number of survivors of each species in your group. (If you have migrated to one or more other groups, report the number of survivors in the last group you joined.)

9. Your teacher will list class totals on the chalkboard. Copy the data in your logbook.

10. Return your fact card to your teacher.

Talk It Over

Work with the other members of your original group to respond to the following items.

1. Why did some small environmental changes cause some animals to die? Did changes affect only one animal or many?

2. What was the reason for flipping a coin to see if your animal survived a harmful change?

Career Links

Agriculture & Natural Resources

U.S. Forest Service

Economic Entomologist

How much do insects damage the environment, and how much does it cost? Economic entomologists study these questions. Insects may damage farms, forests, and livestock. They can also cause disease in humans. Economic entomologists identify insect pests, estimate their numbers, and make predictions about how fast these insect populations may grow and how much damage they might cause. The entomologists also research and carry out various methods of pest control. Some entomologists study how insects develop resistance to pesticides. They must know how pesticides may accidentally harm humans and other animals, including helpful insects. Many entomologists research how to control insect populations without insecticides. For more information, write:

Entomological Society of America
9301 Annapolis Road
Lanham, MD 20706-3115
http://www.entsoc.org

Related occupations: chemical ecologist, pheromone chemist, pesticide specialist, beekeeper, tree expert

3 Did each organism's survival depend on the survival of other species? Explain how this affected the ability of each species to survive.

4 Were there more small changes (affecting a small number of organisms) or large changes (affecting many organisms)? How did this affect the survival of a species?

SPREAD THE WORD

To show what you have discovered about how a species becomes extinct, work with the other members of your group to create a presentation for the rest of the class. A good way to start would be to make a large chart of all the animals in your group. List the name of each species and the resources and environment it needs to survive. Describe how each kind of change affected each of the members of your animal group. Explain which, if any, survived all the changes. Try to explain why you think that species survived. List several professions in which it would be useful to know what causes species to become extinct.

READ ALL ABOUT IT!

The Great Dinosaur Mystery

ONE of the great mysteries of all time is what causes so many species to become extinct. Of the millions and millions of life-forms that have developed on earth, only a small number exist today. Fossils provide evidence of species that flourished and spread across the planet but then suddenly died out. What caused them to become extinct? Dinosaurs are one example of this mystery. Studies of rock formations indicate that by the end of the Paleozoic era (about 245 million years ago), the continents formed one huge landmass. This landmass is called **Pangaea**. The planet was hot and dry. Its shallow inland seas had dried up, and the largest mass extinction in earth's history had occurred. An estimated 95% of all sea life had died out. Only about one in five species of land animals and plants survived.

While the whole world was becoming dryer, the Age of Dinosaurs was just getting started. This age began with what is called the Triassic period. It is not certain whether dinosaurs were warm-blooded or cold-blooded animals, but everything that *is* known about them indicates they were well adapted to earth's new hot, dry environment. A dinosaur's heart and metabolism (internal processes needed to sustain life) were more efficient than those of its reptile ancestors, and its skeleton was better adapted for chasing prey and escaping attackers.

Many useful variations appeared among the earliest dinosaurs. Through the process of natural selection, these useful traits were passed on to offspring. Large communities of different kinds of dinosaurs appeared. Soon dinosaurs of every shape and size were inhabiting every part of Pangaea. Eventually some species of dinosaurs adapted to life in the ocean, while others developed feathers and learned to fly. Thousands of fossils reveal that these reptilelike creatures roamed the planet for nearly 200 million years.

Natural Causes?

Not all dinosaur species were success stories. Some had a short life, existing only one or

During the Mesozoic era (66–245 million years ago), reptiles reached huge sizes and dominated land, sea, and air. The ichthyosaur (A), plesiosaur (B), pterosaur (C), and giant sea turtle (D) were all fish-catching reptiles.

Time Line of Earth's History

Approximately 4.5 billion years ago Earth is formed.

Bacteria appear 3.5 billion years ago.

The first dry land appears 2.7 billion years ago.

two million years. But even those that survived longer did not necessarily extend through the entire dinosaur period. In fact, many species died out many millions of years before dinosaurs as a group became extinct. Because of this, some researchers think that dinosaurs became extinct as part of a natural and recurring cycle: Species appeared and disappeared—as new species developed, older species were often replaced.

Other researchers think there may be more specific reasons to explain why dinosaurs did not survive to the modern period. Fossil evidence shows that the story of the dinosaur follows trends in the climate. When the earth began to warm up, dinosaurs began to thrive. Much of Pangaea centered around the equator. In this tropical climate, plants flourished, increasing both in population and in number of species. Because of all the plants, the earth's atmosphere became rich in oxygen.

Giants Need Lots of Food

Perhaps the dinosaurs that thrived in this environment could not survive as conditions slowly changed. When the continents broke apart and began drifting toward the poles, most climates turned colder. Giant dinosaurs, such as *Tyrannosaurus rex* and *Triceratops*, required lots of food from areas that were getting smaller and becoming more isolated. As food became scarce, conditions grew harsh and many dinosaurs could no longer survive.

It is also possible that colder climates caused either too many male or too many female dinosaurs to be produced. This may have created a mating problem. Today some reptiles lay eggs that are sensitive to temperature. The temperature can affect the sex of the young reptiles in these eggs, sometimes making them either mostly male or mostly female. Dinosaur eggs may have been affected in a similar way.

Fossil records clearly show that massive extinctions have occurred. . . . One of the biggest wipeouts occurred 65 million years ago—the end of the Age of Dinosaurs.

Another theory takes the position that small mammals may have contributed to the extinction of the dinosaurs. These small mammals may have eaten the eggs laid by some species of dinosaurs, reducing the dinosaur population.

These theories do not explain all the evidence that exists. Fossil records clearly show that massive extinctions have occurred—enormous numbers of organisms disappeared in a short time. One of the biggest wipeouts occurred 65 million years ago—the end of the Age of Dinosaurs. For more than a hundred million years, dinosaurs had been extremely successful, but suddenly they disappeared. Not only dinosaurs, but also most marine reptiles (such as sea crocodiles), bony fishes, clams, sponges, snails, and many plants became extinct at this time. For so many species to be wiped out so fast, there must have been

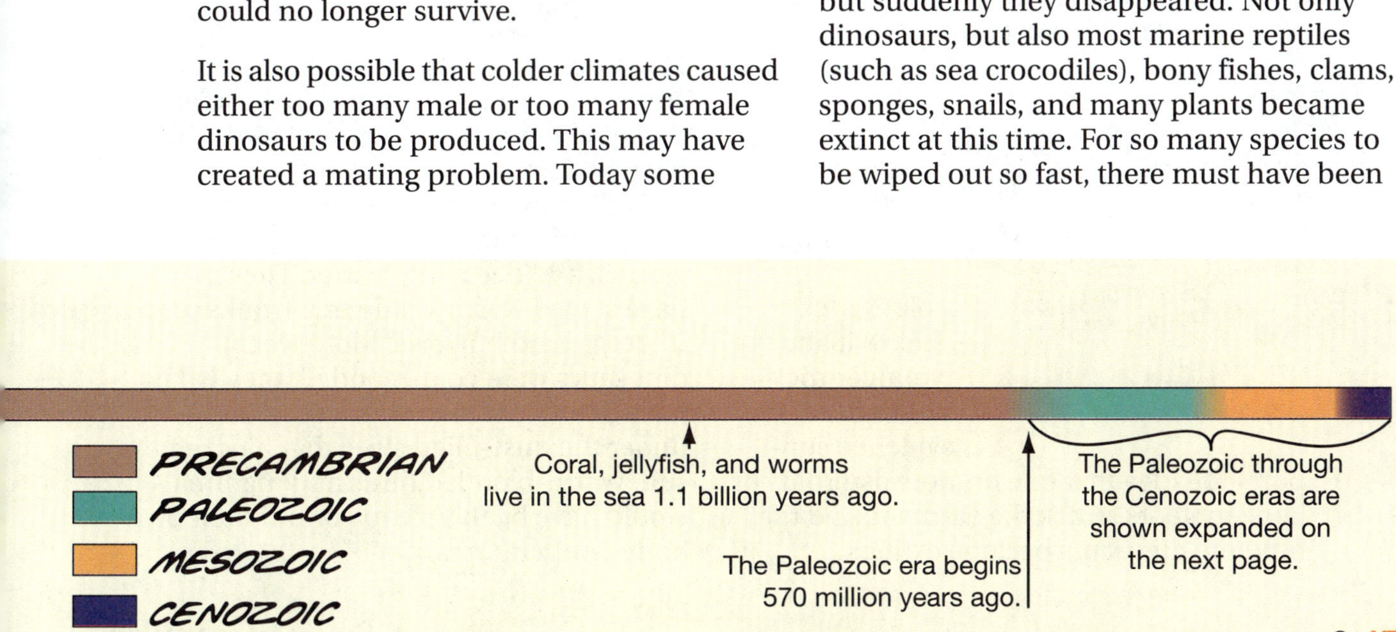

Time Line of Earth's History
(from Paleozoic era to present)

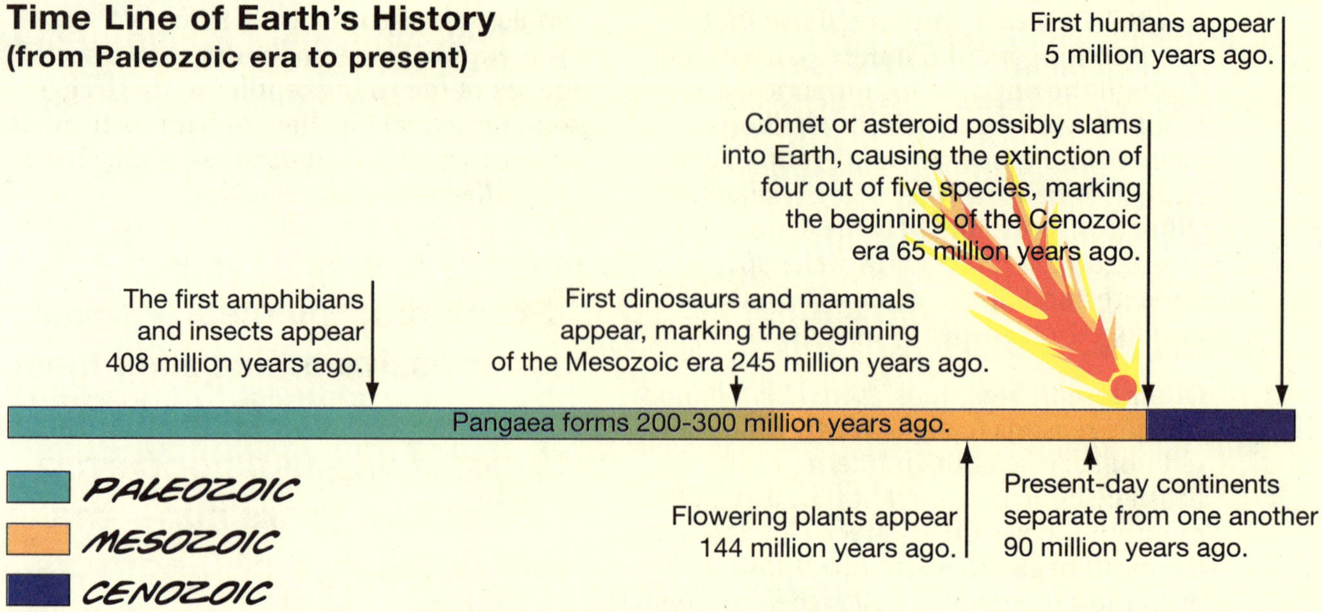

a major catastrophe. Perhaps something happened that changed the climate or the geography of the whole planet. What could have caused such a drastic change?

Cloudy Days

A giant crater about 250 km wide has been detected under the floor of the Caribbean Sea near the Yucatán peninsula in Mexico. The crater lies in a layer of rock whose age indicates that the crater was formed around the time the last dinosaurs died. In addition, a layer of ashes has been found in other rocks of that age all over the world. The chemical composition of this layer of ashes is different from the composition of the older rock below it and the younger rock above it. All this evidence supports the idea that one or more asteroids or large meteors struck the earth around the time the dinosaurs became extinct.

Get This . . .
Only a small number of the millions and millions of life-forms that have evolved on earth exist today.

The force of such an impact would have thrown tremendous amounts of dust and debris into the atmosphere, darkening the skies for months, perhaps even years. Reduced sunlight would have lowered the temperatures and caused many plants to die. With fewer plants and greater competition, many plant eaters would not have obtained enough food to survive. Meat eaters would have had fewer prey to eat, and competition among them would have grown fierce. A chain of events like this one could have caused the dinosaurs and many other life-forms to go extinct.

How could some species of organisms survive such a catastrophe? In a colder, darker climate, mammals and birds would have had an advantage over the dinosaurs in finding food. Perhaps the birds and small furry mammals fed on insects, worms, and decaying matter. They may have stayed warm inside tree trunks or in underground burrows. Moreover, if dinosaurs were cold-blooded, they would have been unable to warm their bodies under the dust-filled skies that blocked the sun. Warm-blooded birds and mammals would have been able to control their own body temperatures.

A Roll of the Dice

Another reason some species survive and others do not has to do with chance. Sheer chance may be as important as adaptation in determining which organisms survive over the long haul. Perhaps some of the dinosaur species that disappeared 65 million years ago would still be alive today, were it not for a cosmic accident, an asteroid from outer space. On the other hand, some species that were rare or endangered when the dinosaurs ruled the world may have enjoyed more success after their powerful competitors disappeared. After the catastrophe that wiped out the dinosaurs, the surviving species of plants and animals went on to populate all corners of the world. The sudden change that occurred 65 million years ago raises some questions: How long will today's species dominate Earth, and could they suffer the same fate as the dinosaurs?

On Your Own

Answer the following questions based on "The Great Dinosaur Mystery."

1. Suppose a type of bacterium that kills many plant species spread over a continent and wiped out most species of plants. Without the plants, most of the animals that consumed them would also die. What do you think would happen to the surviving species on that continent during the next few million years? How is this situation similar to what happened when the dinosaurs appeared?
2. Imagine finding the fossil of a previously unknown dinosaur. How could you find out if that species lived a long time? How could you determine the regions where the species lived?
3. At one time people believed that many dinosaurs died because they fell into tar pits. Like quicksand, tar pits are difficult to escape. Do you think this theory provides a good explanation for the extinction of the dinosaurs? Explain.
4. Mass extinctions may have been caused by catastrophes, such as an asteroid's hitting the earth. How might such events shape the future of life on earth?

INVESTIGATION 6

Survival Strategies

Ever Wonder?

Humans spend a lot of time teaching and taking care of their children. How do other species train their young? For example, do adult cockroaches spend hours every day playing with baby roaches and teaching them how to find crumbs in your cabinets? In this investigation you will play the part of a mosquito, a mouse, or an elephant as you explore the strategies that animals use to make sure their own species survives.

Materials List

- Calculator

Step-by-Step

1 Each member of your group should take the part of either a mosquito, an elephant, or a mouse. Decide who will work on each of these animals.

2 Using the information your teacher has provided, determine how many surviving offspring one female of your species produces in one generation. Record your calculations and answer in your logbook.

3 Assuming that 50% of the surviving young are female, determine how many surviving offspring one female of your species and her descendants will produce in five years. Record your calculations and answers. (**Math tip**: When you do calculations with more than two numbers, write out the whole calculation, using both numbers and names in fractional form. Then cross-cancel names that appear in both a numerator and a denominator.)

For example: $$\frac{12 \text{ mice}}{21 \cancel{\text{days}}} \times \frac{30 \cancel{\text{days}}}{1 \text{ month}} = \frac{17 \text{ mice}}{1 \text{ month}}$$

4 Compare your answers with those of your group members. Record their results in your logbook.

Talk It Over

Work with the other students in your group to answer these questions:

1 What would account for the differences among these three species in the number of young that survive long enough to reproduce?

2 Consider reproduction in other animals, such as cats, flies, fish, birds, and humans. Is there a relationship between the intelligence level of each species and the amount of time and energy spent on caring for its young? Explain.

3 Think about the three kinds of animals that you used in the investigation and the other animals that you considered in the preceding question. Do you see any relationship between the number of offspring and the resources devoted to caring for them? Explain.

Spread the Word

Work with other members of your group to give a presentation to the rest of the class. Your goal is to show what you have learned about the connection between the number of offspring that a species produces, the number of offspring that survive, and the amount of care that parents devote to their offspring. To help you prepare, use an encyclopedia or animal guide to find information about various animals. Research their breeding habits, and find out how many offspring they produce. You could also make a bar graph that illustrates your information: For each animal, graph the number of offspring. Make a poster-sized version of the graph so that the class can see it more easily, and point out any trends that the graph reveals.

List some jobs and other situations in which people would need to know about the reproductive strategies of animals.

It Takes All Kinds of Parents

Does a mountain gorilla have anything in common with a squid? Mountain gorillas are among the world's most intelligent mammals, and squid may be the smartest invertebrates (animals without backbones). But when it comes to reproduction, these two species are as different as they look.

The opalescent squid is a small, shimmering creature that lives less than a year. Every summer thousands of squid gather at a spawning site, where they court and mate. The females attach their white egg cases to the seafloor. Shortly afterward, the adults die. The hatchlings will never know their parents, and they will have many predators. However, with thousands of eggs laid at once, enough are likely to survive to carry on the species.

The mountain gorilla is a much different kind of parent. Gorillas can live more than 35 years. Females start breeding when they are about 10 years old; males, somewhat later. Family groups consist of a mature male leader called a silverback, two or three females, and several young. A newly mature male leaves the group to travel alone or with other males until he can attract a female and begin a new family. A female gorilla will have only two or three offspring during her lifetime. Her babies will nurse until the age of two, and they will sleep in her nest for another year after that.

A silverback is very protective of his family. This normally gentle animal will die trying to protect his young. The death of a silverback can splinter the group. Although the mountain gorilla's only enemy was once the leopard, it is now threatened by human poachers. The poachers invade the two protected areas in Africa where mountain gorillas live.

Only about 650 of these animals still survive. Once some humans became their enemies, the gorillas' low birthrate, the amount of time it takes for their young to mature, and their stable family structure may have worked against their survival.

Read All about It!

Finding Ways to Survive

It's a jungle out there. Many animal species are constantly in danger, either from predators or from harsh environments. They also have to compete with members of their own species. With all the problems and dangers they face, how do the smaller, weaker species survive?

Animals have found many ways to adapt their reproductive strategies to the game of survival. If an animal can avoid being eaten, it can reproduce. Amphibians, fish, and insects produce hundreds or thousands of offspring at one time. They need to do this because they do not protect their young and because few offspring survive to reproduce.

The strategy works. If you don't think it does, just look at how many bugs exist. Insects are one of the most successful groups of animals in the world. They first appeared more than 300 million years ago, and today they outnumber all other animals combined. They have invaded every part of the planet—land, air, and water.

Big Families

How have insects managed to do so well? Besides laying hundreds or thousands of eggs in a single breeding season, each generation of insects matures quickly. Soon the new generation is producing the next generation. Their short generation time allows mutations and adaptations to occur quickly. As a result, insects have become highly diverse and specialized. For example, consider the different kinds of mouth parts that insects have developed (see Figure 8).

The mouth of a female mosquito enables her to suck blood, which she must have before laying her eggs. She gets blood by inserting a delicate tube into her victim's skin and then drawing blood through the tube. But the male mosquito, who has only simple mouth parts, must survive on nectar and water. Houseflies have a spongelike mouth that they use to soak up liquids. With so many mouths to feed, if all insects competed for the same food supply, they would have even more trouble surviving.

Some insects solved the problem of competition for food between young and old by dividing their lifetime into two parts. For

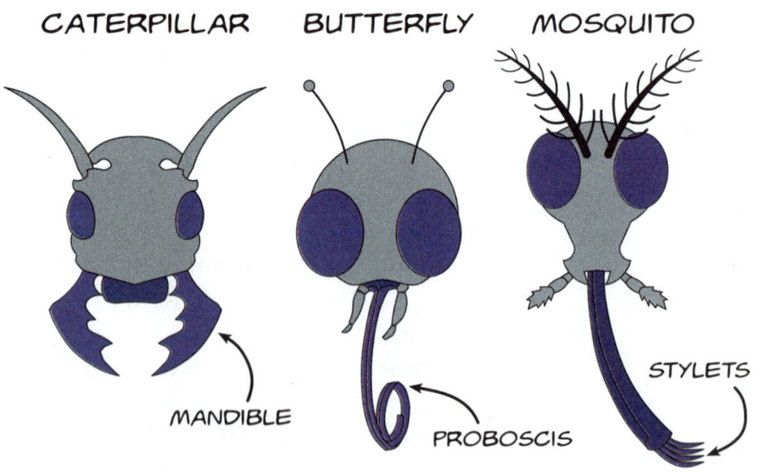

Figure 8: Insect mouth types
Note: Caterpillars have mouths that are specialized for chewing, butterflies for sucking, and mosquitoes for piercing and sucking.

If an animal can avoid being eaten, it can reproduce.

instance, a caterpillar uses its strong mouth parts to chew leaves, but later the caterpillar turns into a butterfly, which has a long, coiled mouth tube for sucking nectar from flowers. With their specialized mouth parts, all these insects can coexist.

Reptiles lay fewer eggs than insects, fish, and amphibians. Some reptiles guard their offspring until they can take care of themselves. The Nile crocodile carries her young in a special pouch in her mouth. Both parents watch the offspring for several months as the young crocodiles learn to hunt small fish and frogs. On the other hand, some reptiles, such as sea turtles, do what insects do and simply leave their eggs shortly after laying them.

Bringing Up Baby

Animals that produce only a few offspring spend a lot of time caring for their young. Most birds and mammals care for their offspring until the young have developed physically and have learned the skills necessary for defense and feeding. Depending on the species, this period may last weeks, months, or years. Most birds and mammals have a long training period between the time they are born and the time they mate. Although a newborn female elephant joins the herd only half an hour after birth, she will not start reproducing until she is 15 years old. The elephant must first learn how to use its large brain and grasping trunk to obtain food, shelter, and safety before it is ready to teach these skills to its own young. This is the best way to make sure the process will be repeated in the next generation.

The two groups of animals that you examined during Investigation 6—insects and **vertebrates** (animals with backbones, such as fish, amphibians, reptiles, birds, and mammals)—have nearly opposite reproductive styles, but they do have one thing in common. Members of both groups have jointed limbs, which are found only in these groups and in relatives of the insects, such as crabs, spiders, and centipedes.

Jointed limbs—legs, flippers, wings, and arms—have contributed greatly to the success of these animals by allowing them to travel over land and through water and air. These limbs also enable them to handle objects in their environment. They can use their limbs to build nests, catch and hold prey, defend themselves, and collect objects. Vertebrates have a large brain that controls their limbs, allowing the animals to learn new behaviors and to use them in new or changing environments. Insects have smaller, simpler brains. They must rely on genetically programmed instructions for their behavior. Nevertheless, some insects, such as ants and bees, which live in large groups, are able to cooperate to build nests, collect and distribute food, and defend the group.

Queen Rat

It is wrinkled, bucktoothed, and almost bald. As one scientist said, it looks like a hot dog that was left in the microwave too long. As if that weren't weird enough, it's a mammal that behaves like an insect. What is it? It's a naked mole-rat.

This African rodent is the only mammal known to reproduce the same way bees and ants do. Colonies of naked mole-rats are composed of the offspring of a single queen who mates with a few males. Like insect queens, the queen mole-rat is much larger than her relatives. She will threaten other females, and she emits a chemical that interferes with rivals' ability to breed.

The males are all capable of mating, but no more than three in a colony get to do so. The vast majority of naked mole-rats have specific tasks to perform for the good of the colony, including tending to the queen's offspring.

Individuals in naked-mole-rat colonies are nearly identical genetically. There seems to be no biological need for individuals to pass on their own genes—others in the colony do it for them. The system works well for this highly social animal, which spends its life in the underground colony. They build and maintain tunnels in dry regions where the ground is as hard as rock. To accomplish this, they all must work together. Naked mole-rats have another characteristic that makes them depend on one another—they cannot regulate their body temperature. Because of this, they must often huddle together to keep warm.

On Your Own

Respond to the following items based on "Finding Ways to Survive."

1. It used to be common for zoo workers to remove newborn animals from their parents. Separation of young animals from their parents can have dramatic effects on the social behavior of the young. Describe different kinds of behavior in which these effects can be observed.
2. How is play important for young animals?
3. Suppose you find a beetle that has a mouth with large, powerful pincers. What kind of food do you think such a beetle would eat?
4. How would you explain the fact that many vertebrates live long after giving birth but few insects do?

Drifting Apart

Ever Wonder?

Think back to the first time you visited a zoo. Were you amazed to see so many different kinds of animals from other parts of the world? Did you wonder why some types of animals don't live in this part of the world? No monkeys hang out in your local park, but plenty of squirrels probably do. Both monkeys and squirrels are furry mammals that live in trees, so why do squirrels live here but monkeys live in Africa? If you take a ride in the country, you might see deer grazing in a field but you won't see any kangaroos or koalas—unless you're in Australia. And you will never see anything as strange as a three-toed sloth climbing a telephone pole in an alley.

In this investigation you will discover the reasons different types of animals live in different regions of the world.

Materials List

- Fact cards

Step-by-Step

1. Examine the fact card you have received. In your logbook, sketch the organism shown on the card. Write the name of this species under your drawing.
2. In your logbook, record what the card says about this species.
3. Look at the cards that the other members of your group received. Work with them to discuss what kind of environment each species needs.
4. After your group reaches a decision on all four species, record in your logbook the conclusion about the species on your card.
5. Next, imagine that your group's four species are living on a huge continent when a major geographical event occurs: The large continent breaks up into

smaller continents that slowly drift apart. Many changes occur. For example, some places that used to be dry inland regions suddenly have seacoasts; some continents drift closer to the poles; some move toward the equator. Basically, four kinds of changes occur: Each new continent becomes wetter or drier, hotter or colder.

6 Discuss with the other members of your group how each species might change and diversify in order to succeed. Consider these questions: Would the species survive in a changing environment? How might its descendants adapt to a changing environment?

7 After your group completes its discussion of all four species, record in your logbook your conclusions about your own species.

8 Draw a sketch of how your species' descendants might look after many generations of adaptation to each of the four changed environments.

9 Write a few sentences to go with each of your sketches. Describe the special adaptations that would allow each descendant to succeed. Explain how these features would help it survive.

10 When all members of your group are finished with Step 9, leave your group and join the members of other groups who were assigned the same species that you were.

11 Compare your sketches and ideas with those of other students who worked with your species. In your logbook, record a brief description of each person's ideas about how the species might change in adapting to different environmental conditions.

TALK IT OVER

Work with the other members of your new group—the students who worked on the same species that you did—to answer the following questions.

1 What kinds of characteristics helped your species adapt to its original environment? Explain.

2 If part of its environment changes, would it be *necessary* for a species to change in order to survive? If part of its environment changes, would it be *likely* that the species would change? Explain your answers.

3 Did every student in your new group come up with the same solutions to living in each of the four changed environments? Is it likely that a species would produce more than one type of descendant that would be adapted to the same environmental change? Explain.

4 Do the results of the investigation make you think that you would find the same species in the deserts of the southwestern United States, the Sahara in North Africa, and the deserts of central Asia? Explain your ideas.

5 How could you use maps showing where various species live to help you identify the continents that were the first ones to separate from the others?

Spread the Word

Work with the members of your original group to create a presentation that shows the rest of the class what you have learned from this investigation.

An effective presentation will show how continental drift and changes in climate may have determined where different kinds of species live today. You could use enlargements of your drawings to show the class how you developed your ideas. Give your reasons for thinking that changes in the environment may cause the changes you predicted. You could also use a map of the world to show where certain plants or animals thrive today, and you could explain what this can reveal about the order in which various landmasses separated.

Describe several professions in which knowledge about how the environment affects species might be useful. (Hint: Think about workers who would need to know where various species are found.)

Read All about It!

The Breakup

WHY do wild kangaroos and koalas live only in Australia? If pigeons can be found all over the world, why do most species of mammals live only in certain areas?

To answer these questions, you must peer into the distant past, hundreds of millions of years ago, before the first reptiles appeared. All the land on earth formed one giant continent called Pangaea. Plants and animals could spread easily across that single landmass. Later, during the time of the dinosaurs, Pangaea's crust began to crack because of the pressure of heat deep inside the earth. As the land separated (see Figure 9), oceans filled the gaps between the new, smaller landmasses, which eventually became the continents that exist today.

At first many species lived on most of the new continents. For example, dragonflies and conifers (trees with needle-shaped leaves) were common on Pangaea. Researchers know this because fossils of dragonflies and conifers have been found all over the world. Moreover, both these groups still live all over the world. Today's conifers and dragonflies are quite different from one another, but they all came from the original species that lived on Pangaea. Members of the original species became isolated from one another because of **continental drift**, the slow movement of continents that has been occurring for hundreds of millions of years.

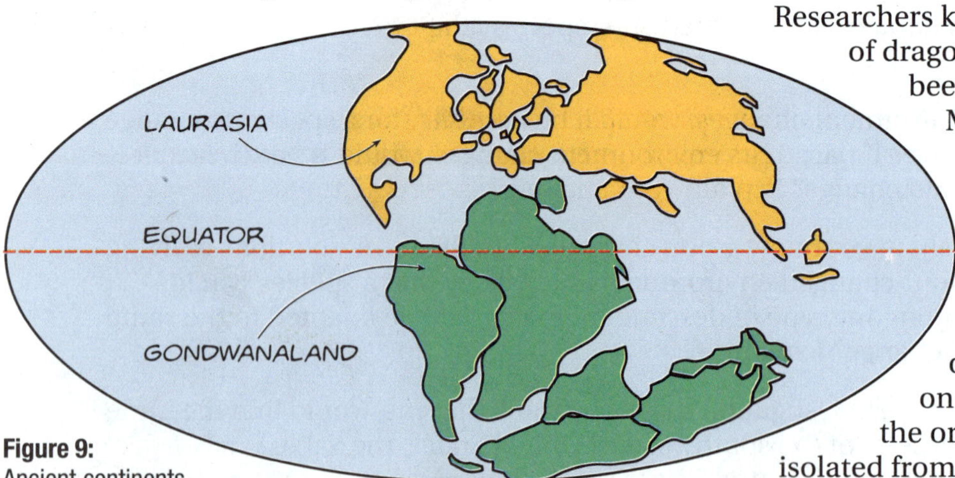

Figure 9:
Ancient continents
Note: During the late Mesozoic era (150–200 million years ago), Pangaea broke up into two large continents, Laurasia in the north and Gondwanaland in the south.

Rearranging the World

After Pangaea broke apart, some continents drifted toward the poles. Northern Eurasia and much of North America cooled down to ice-age temperatures. Northern South America, Africa, and much of Asia grew warm and tropical because of the constant moisture from ocean waters warmed by the sun. Mountain ranges, such as the Andes and Rockies, rose up, blocking the movement of moist sea air and rain clouds. This created new deserts. In some places, including Australia, forests gave way to vast semiarid grasslands.

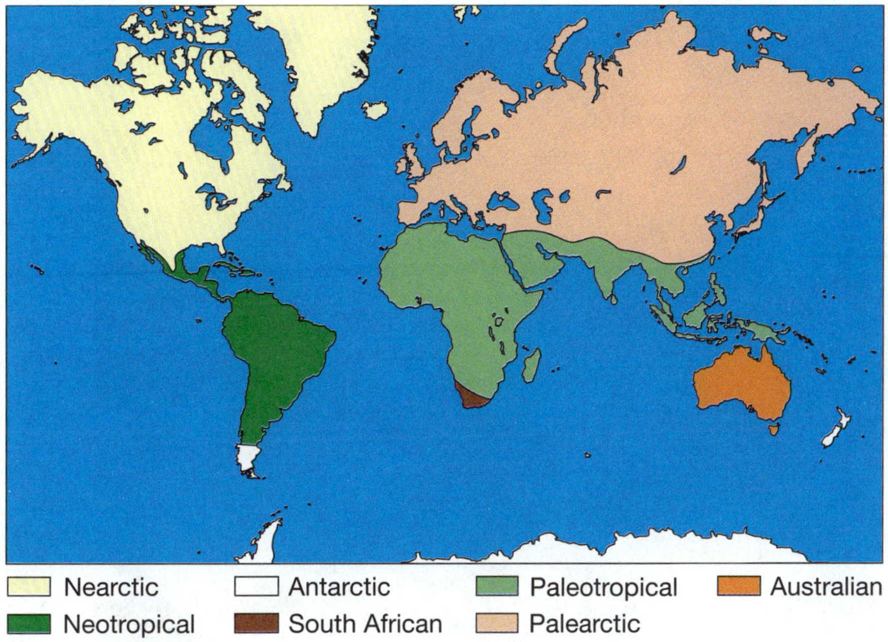

Figure 10: Biogeographical realms
Note: Similar environments in different realms are inhabited by different species.

Today the continents are divided into seven **biogeographical realms** by barriers such as oceans, mountain ranges, and deserts (see Figure 10). The boundaries of the realms are based on their animal and plant species. Different realms may contain similar ecosystems made up of different species. For example, grasslands similar to the American prairies occur in West Africa and Tibet, but different kinds of grasses and animals live in each place. If you compare the bison that lives in America with the yak that lives in Tibet, you will see that these two species are closely related. However, their differences have increased since they were separated by continental drift.

Marsupials Down Under

Australia was the first continent to separate from Pangaea. This helps to explain why Australia has several unique species of wildlife. Australian mammals, such as kangaroos, wallabies, and koalas, are marsupials. A **marsupial** is a primitive mammal that gives birth to undeveloped offspring, which the mother carries in a pouch. Most other mammals are **placentals**, which develop entirely inside their mothers' wombs before birth.

If you compare the bison that lives in America with the yak that lives in Tibet, you will see that these two species are closely related.

After Australia broke away from Pangaea, the marsupial species still on Pangaea gave rise to placentals. Eventually the placentals replaced most marsupials on Pangaea. In North America, only one marsupial species, the opossum, has survived. But Australia has nearly a hundred different marsupial species. What caused this? The evidence points to Australia's early separation from Pangaea. Because Australia became isolated before the placental mammals appeared—and because placentals were unable to cross the water to get there—the marsupials on Australia were protected from competition with the more advanced mammals.

The cuscus is an Australian marsupial that has adapted to fill a niche occupied elsewhere by monkeys. Note its long tail and its generally monkeylike appearance.

Over millions of years, Australia's marsupials have diversified greatly. Today these animals use the environment in the same ways that placentals and other animals do—they have carved out different niches for themselves. For example, the honey phalanger, a squirrel-like marsupial, sucks nectar from flowers the way a hummingbird does. The kangaroo rat hops, hides, and grazes like a rabbit. The cuscus lives in trees and uses its grasping tail like a monkey. These marsupials are related to one another, not to the birds, rabbits, and monkeys whom they resemble in certain ways. By using the environment in the same way, different species, such as the kangaroo rat and the rabbit, have developed similar types of bodies. The development of similar adaptations in unrelated species is called **convergence**.

On Your Own

Answer the following questions based on "The Breakup."

1. If you traveled to another planet that had separate continents, would you expect most of the plant and animal life to be the same on each continent? Explain.
2. Visitors to the islands of New Zealand, off the coast of Australia, are not allowed to bring their pet cats or dogs with them. What might be the reason for this law?
3. Suppose nectar-drinking birds that are not related to North American hummingbirds live in another part of the world. If these birds have long, thin beaks that are similar to hummingbirds' beaks, would you expect to find similar flowers in both places?

Section Wrap-up

The Story of Life on Earth

Life today is very diverse. Just think of all the different kinds of insects you see flying or crawling around—bees, crickets, flies, mosquitoes, grasshoppers, walking sticks. . . . The list runs on and on. There may be two million species of insects—no one has ever counted them all. Then there are all the other species of animals, plants, fungi, bacteria, and other microscopic organisms. Yet despite the tremendous number of species, only a small number have been around longer than a few million years.

Fossil and genetic evidence indicates that the great diversity in today's species is the result of several hundred million years of changes to the original life-forms. One of the most significant changes in species occurred when land formed. Ocean-dwelling plants began floating ashore. The ones that could survive direct sunlight and dry air crowded the beaches. Over millions of years, some of these plants developed stems, leaves, and roots that enabled them to support their growth, obtain and transport nutrients, and avoid drying out. Mosses, horsetails, and ferns were among these successful species, and they spread inland.

Changing Times

Fossil evidence also shows that a few species of marine animals also managed to survive the harsh conditions on land. Some fish trapped in small pools during droughts were able to gulp air and survive until the next rain. Others found their fins enabled them to move to new bodies of water. After a few million years of natural selection, the animals that could survive on land had new variations, such as jointed legs, larger brains, and lungs. Each of these variations represented a successful adaptation to land.

More land emerged from the sea, and new environments developed. With every environmental change, new variations appeared in organisms. Some of these variations were helpful adaptations that enabled the species to survive. For example, the first froglike amphibians lacked the protective armor of arthropods, but they could move very fast. They were successful as a species, but they had to remain near water in order to reproduce. Over the years, variations among these amphibians produced some that had dry, waterproof skins and egg coverings that made reproduction possible away from water. These new animals—reptiles—used this advantage to move away from the water and roam the land.

World in Upheaval

As more species of land-dwelling plants and animals developed, the first landmasses were being carried along by continental drift. Evidence from fossils and patterns of rock types suggests that eventually these land areas crashed together, forming a

Get This . . .

Sheer chance may be as important as adaptation in determining which organisms survive over long periods of time.

single huge landmass called Pangaea (see Figure 11). The collisions caused vast changes in the environment. Earthquakes and volcanic eruptions racked the entire planet. The earthquakes and volcanoes were much stronger than today's. They changed the landscape dramatically, pushing up mountains and drying up inland seas and rivers. By the time these massive changes ended, an estimated 245 million years ago, only one in five species was still alive.

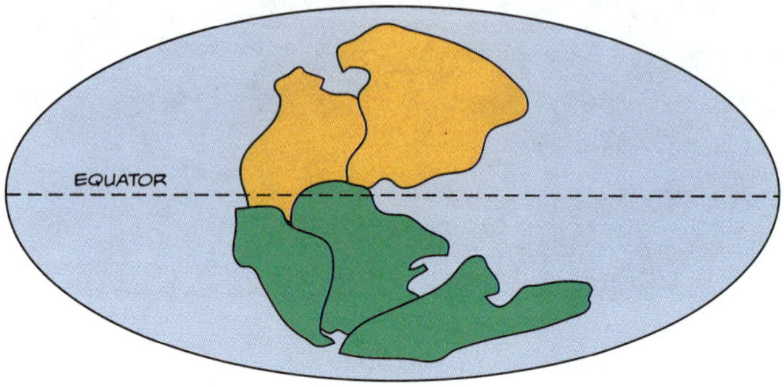

Figure 11: Pangaea
Note: The continents are thought to have collided, forming a single large mass, during the Paleozoic era (200–300 million years ago).

The sudden mass extinction made room for new plant and animal species. Fossils from the period following this extinction indicate that tremendous adaptive radiation occurred among the insects and reptiles that survived. New kinds of plants emerged, including trees that could live away from surface water. First conifers and then flowering plants provided new homes and foods for insects, which developed various types of wings and legs as they competed for these new resources. In turn, these insects provided food for a new group of animals, the small, warm-blooded ancestors of mammals.

Day of the Dinosaur

Another type of reptile, the dinosaur, also developed at this time. Birdlike hips enabled some dinosaurs to stand upright and run quickly. These traits gave them a big advantage over other animals, which became easy prey. Many different types of dinosaurs developed, and each species found its own niche. Some lived on land, some in the sea, and others flew. Some were predators. Others fed on plants. Some were small scavengers, while others were the size of small buildings. Some moved on all fours, and others walked on their hind legs. For an estimated 165 million years, the many species of dinosaurs ruled the earth.

During the reign of the dinosaurs, Pangaea broke up and the continents began drifting apart. Many species became separated from their relatives by widening oceans. As environments slowly changed, species that could not adapt to the new conditions—hotter or colder climates, wetter or drier seasons, and new foods or predators—died out. But other species developed variations that helped them adapt to a new environment. These successful species took over the roles that the extinct species had played.

Most insect species succeeded by having large numbers of offspring—some offspring were always sure to survive. Birds and mammals followed a different strategy. They had few offspring, but the parents took care of the young for extended periods of time. When these offspring finally left the nest, they were older and knew how to survive.

Still Drifting

Because of continental drift and the environmental changes it caused, species whose members no longer lived on the same landmass began to diverge. Today related species on separate continents may look quite different from one another, even though they have a common ancestor. For example, the closest relative of North America's poison ivy is the African cashew tree. As the environments of these two continents changed, the process of natural selection favored a different adaptation for the common ancestor in each place.

The incredible variety of life that you see today is the result of hundreds of millions of years of changes and thousands of dead ends. These changes are still occurring. The continents continue to drift . . .

On the other hand, some unrelated species have converged, developing similar adaptations in response to natural-selection pressures. For example, bats have come to resemble birds, and some Australian marsupials resemble monkeys or rabbits in their traits and behaviors.

The incredible variety of life that you see today is the result of hundreds of millions of years of changes and thousands of dead ends. These changes are still occurring. The continents continue to drift, and species continue to adapt to their changing environments. How will the world and the species that live on it look in the future?

On Your Own

Respond to the following items based on "The Story of Life on Earth."

1. Squirrels often bury nuts to maintain a food supply they can use later. Explain how this habit may have been a factor in the development of plants that produce nuts (a form of seed).

2. Would continental drift affect bird species less than mammals? Would this always be the case? Explain your ideas.

3. Imagine you are an archaeologist examining the rock layers under a desert. Suppose you discover that fossils of a particular species of lizard are not present in rocks younger than a certain age. Could you conclude that the species died out? If so, how could you tell whether it died in a mass extinction?

4. What if you were exploring a planet that was completely covered by water teeming with life? If you took samples from the rocks on the ocean floor and found several fossils of flowering plants with roots and leaves, what would you conclude about the history of this planet and its life-forms?

Section 4

THE FUTURE OF LIFE AND LAND

Do you own a dog? Has it ever had fleas? If so, you probably know that fleas are hard to eliminate completely. Even when you apply a spray or powder with chemicals that can kill fleas, the tiny itchy bugs survive. They hop off the dog onto the carpet or furniture. They wait until the poisonous chemicals are gone, and then they hop back onto the dog.

What would happen to the fleas if suddenly they had no safe place to go? A changing environment can be deadly to any organism. You have already seen how environmental changes can wipe out entire species and give rise to new life. In Section 4 you will examine the kinds of environmental changes that are occurring today, and you will discover how they might affect the future of life on earth. Only those species with variations that can adapt to the new conditions will survive.

As you watch the next video segment, think about how today's environment is changing. How will these changes affect life? Are humans and other organisms like fleas with no safe place to jump?

FREEZE FRAME

After viewing the video segment, work with the other students in your class to respond to these on-screen challenges:

1. How does human population growth affect the environment?
2. What factors are causing the environment to change today?

Something's in the Air

EVER WONDER?

Have you ever noticed a dandelion or some other plant growing in the middle of a sidewalk or a road? Maybe you have seen pictures of plants in a desert or an arctic waste. Did you wonder how these plants could survive in these places? How can there be so many different kinds of plants living in so many different places? What role do plants play in the environment?

In this investigation you will discover several ways that plants contribute to their environments. As you explore the links between plants and other forms of life, you will come to understand one reason for the diversity of life in today's world.

MATERIALS LIST

- 2 large, clear plastic containers with glass tubing inserted in two-holed lids
- 2 short pieces of rubber tubing
- 4 long pieces of rubber tubing
- 4 tubing clamps
- 2 flasks (500 mL)
- Baking soda solution
- Graduate cylinder (100 mL)
- 2 beakers (150 mL)
- 2 one-holed rubber stoppers with glass tubing
- 2 ringstands with clamps
- Vinegar
- Small potted plant
- Lamp with fluorescent bulb
- 2 wooden splints
- Matches
- Tongs
- Safety goggles

STEP-BY-STEP

Day 1

1. Put on your safety goggles.
2. Insert a one-holed rubber stopper with glass tubing into the mouth of each flask.

3 Stabilize the flasks by securing each of them to a ringstand with a clamp.

4 Water the plant.

5 Carefully lower the plant into one of the plastic containers. Make sure you do not crush the stems or leaves. Position the plant upright.

6 Wet one end of each of your six rubber tubes (two short, four long).

7 Take one container lid with two glass tubes through it (see Figure 12). Slide the wet end of one long rubber tube onto the top of one of the glass tubes.

8 Slide the wet end of a second long rubber tube onto the bottom of the same glass tube.

9 Slide the wet end of a short rubber tube onto the top of the other glass tube.

10 Repeat Steps 7–9 with the second lid.

11 Place a tubing clamp on each of the rubber tubes that you have attached to the top of each lid. Leave the clamps open.

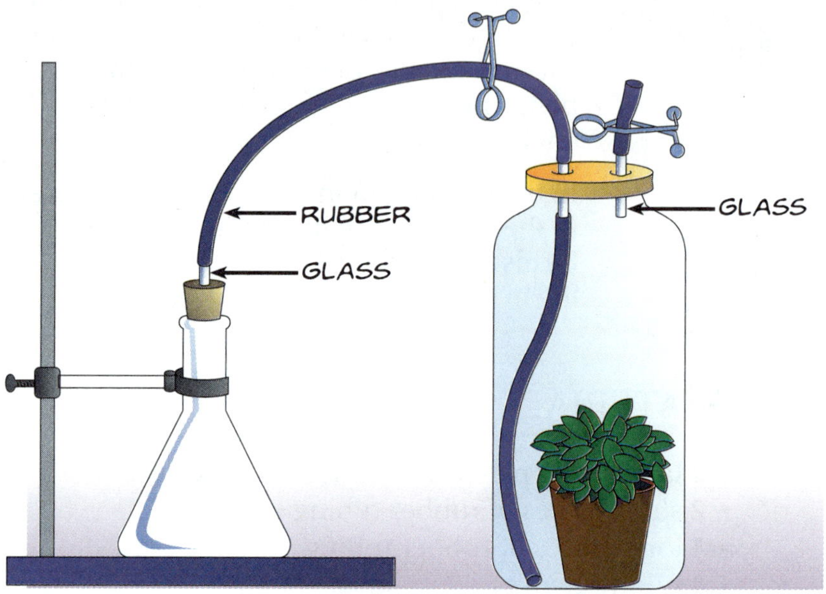

Figure 12: Equipment setup for Investigation 8
Note: Leave both clamps open until Step 20.

12 Wet the free end of the long rubber tube that is attached to the top of one lid. Carefully attach this end to the glass tube in one flask (see Figure 12).

13 Repeat Step 12 with the other lid and flask.

14 Make sure both containers and both flasks have tight seals around all tubes.

15 Carefully attach the lids to the plastic containers. Make sure the rubber tube that goes inside the container with the plant does not rest on top of the plant. (The end of this tube should be near the bottom of the container.)

16 Measure 100 mL of vinegar in the graduate cylinder. Pour the vinegar into one of the beakers.

17 Repeat Step 16 with the other beaker.

18 Remove the stopper from each flask. Pour the vinegar from one beaker into one flask, and immediately reinsert the stopper. Then pour the vinegar from the second beaker into the second flask, and immediately reinsert the stopper.

19 Observe for three minutes. Describe in your logbook what you see in the flasks.

20 When the reaction in each flask ends, close both hose clamps tightly. Carefully separate the rubber tubing from the glass tubing in the flask stopper.

21 Place the lamp next to the two plastic containers, and turn it on. Be sure light strikes the side of each container. (Both containers must be exposed to the same amount of light.)

22 Leave the lamp turned on overnight. Clean up your work area, and return all other materials.

Day 2

1 Put on your safety goggles.

2 Hold one splint with tongs.

3 Light the splint with a match. Let it burn for only two or three seconds, and then blow it out so that its embers continue to glow without a flame.

4 Have one group member—someone who is not holding the splint—slowly and gently remove the lid from the plastic container without the plant. The group member holding the splint should then insert it into the container. Observe. Describe in your logbook what occurs.

5 Use tongs to remove the splint. Hold the splint under running water to extinguish any embers that are still glowing.

6 Repeat Steps 2–5 with the container in which you put the plant.

7 Clean up your work area, and return materials to your teacher.

TALK IT OVER

Work with other members of your group to answer the following questions.

1 What resulted from the reaction between the baking soda and the vinegar? Why was it important for this reaction to occur before you sealed the containers?

2 Why was it important to use two containers?

3 What effect did the light have? Did it have the same effect on both containers? Explain your answers.

4 What was the reason to wait until the next day to test the containers with splints?

5 What happened to each splint when you put them into the containers? What do the results tell you about plants?

6 What substance that is needed to support a fire did the plant add to the air in its container? Do other living things require this substance? Explain your thinking.

7 Would the results have been different if you had used a larger plant? Describe any differences that you think would have occurred, and give your reasons.

8 If all plants exchange gases in this way, how do they affect the atmosphere?

9 What would happen if all the plants on earth disappeared?

Spread the Word

Work with the others in your group to share what you discovered in this investigation with the rest of the class. You could start by describing how you set up your equipment and by explaining what happened in each container. Be sure to give the reasons that you think the splints reacted as they did. Also explain what you think caused each splint to react the way it did.

From what you have learned in this experiment, describe how plants affect the atmosphere and other living things. List several professions in which this information would be important.

Read All about It!

The Web of Life

EVERY kind of organism on earth plays an important role in the game of life. As you discovered in Investigation 8, plants remove carbon dioxide from the air. This is important because CO_2 can be deadly to many organisms, including you: If the air contains too much CO_2, it can dilute the oxygen you need to breathe. Plants also produce oxygen. Animals and other living things need oxygen to release the energy in their food. Through the process of photosynthesis, plants help to maintain a delicate balance between oxygen and carbon dioxide in the atmosphere (see Figure 13). Much of the life that exists on earth today depends on this balance.

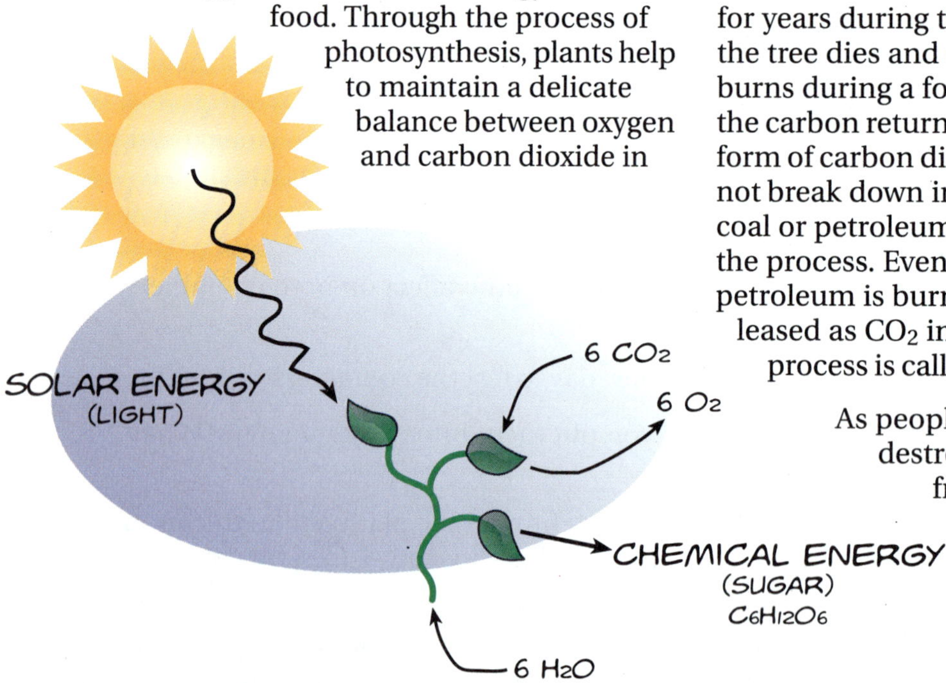

Figure 13: Photosynthesis

Note: Plants absorb water (H_2O) through their roots and carbon dioxide (CO_2) through their leaves. They use the carbon atoms in six CO_2 molecules to form a six-carbon sugar molecule ($C_6H_{12}O_6$) and six molecules of oxygen (O_2). Solar energy is stored in the chemical bonds of the sugar.

Recycling the Natural Way

Carbon dioxide and oxygen cycle back and forth between organisms and their environment. For example, some of the carbon in the CO_2 that is absorbed by a tree helps to make wood. This carbon may sit in the wood for years during the life of the tree. When the tree dies and decomposes—or when it burns during a forest fire or in a fireplace—the carbon returns to the atmosphere in the form of carbon dioxide gas. If the wood does not break down in some way, it can become coal or petroleum. But this merely delays the process. Eventually, when the coal or petroleum is burned, the carbon will be released as CO_2 into the atmosphere. This process is called the carbon cycle.

As people clear forests, they destroy plants that remove CO_2 from the air. This causes the amount of CO_2 in the air to increase. Clearing land by burning dry vegetation releases more CO_2. As CO_2 builds up in the atmosphere, it has a major effect on the earth's climate.

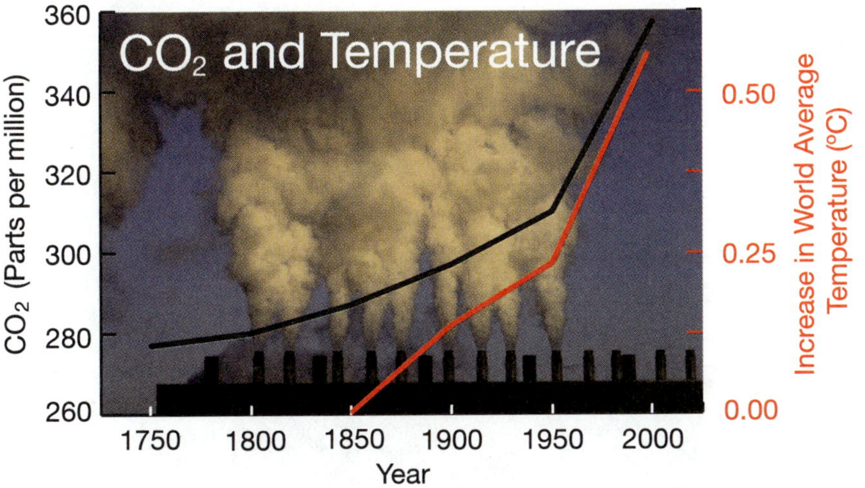

Figure 14: Changes in atmospheric CO_2 and world average temperature
Note: Increasing CO_2 emissions from agriculture and industry may be warming the atmosphere. Most of the increase is caused by the burning of forests to clear land for farms and by the burning of agricultural wastes such as stems, straw, and weeds.

To help you understand how CO_2 affects climate, recall what you observed in the last demonstration. You saw that a bright light warms a beaker containing a large amount of CO_2 faster than a beaker of ordinary air. How does this happen? As the lamp's infrared rays heated the black paper, the pieces of black paper warmed the beakers. At the same time, the beakers lost heat by emitting infrared rays in all directions. The CO_2 molecules in the beakers absorbed some of these rays. The CO_2 molecules also reemitted the infrared rays in all directions. Some of these rays were reabsorbed by the beakers and the black paper. The beaker that contained a large amount of CO_2 warmed up more quickly because of the ability of CO_2 to recycle heat in this way. Similarly, CO_2 in the atmosphere helps the earth retain the heat of the sun's infrared rays. The more CO_2 that the atmosphere contains, the warmer the climate grows.

For hundreds of millions of years . . . a delicate balance existed between the amounts of oxygen and carbon dioxide in the atmosphere.

For hundreds of millions of years—ever since photosynthetic plants appeared—a delicate balance existed between the amounts of oxygen and carbon dioxide in the atmosphere. (Researchers can determine the past composition of the atmosphere by analyzing rocks and fossils.) During the past hundred years, however, the balance has been changing. Since the late 1800s, the amount of CO_2 in the atmosphere has increased greatly. Records from all over the world show that average temperatures from year to year have tended to rise during the past century (see Figure 14). This trend is called global warming or the **greenhouse effect**.

Population Explosion

What is causing these changes? Ever since the industrial revolution, which began in the late eighteenth century, the human race has been surging in numbers (see Figure 15). To grow food for the growing population, people have been burning grasslands and cutting down forests. This has added CO_2 to the air and destroyed the

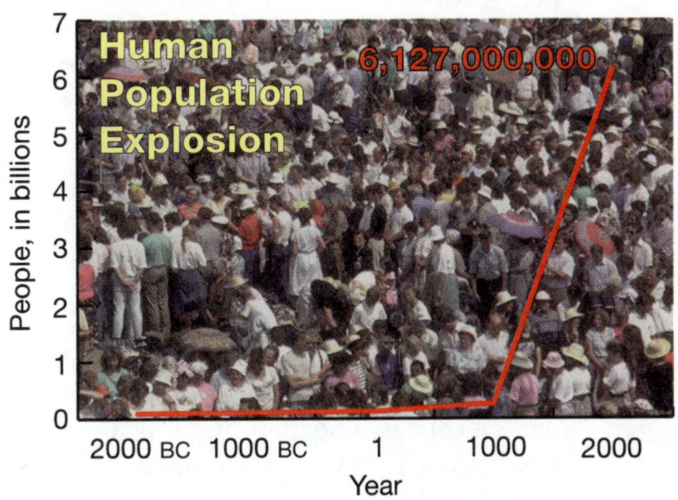

Figure 15: Worldwide human population growth

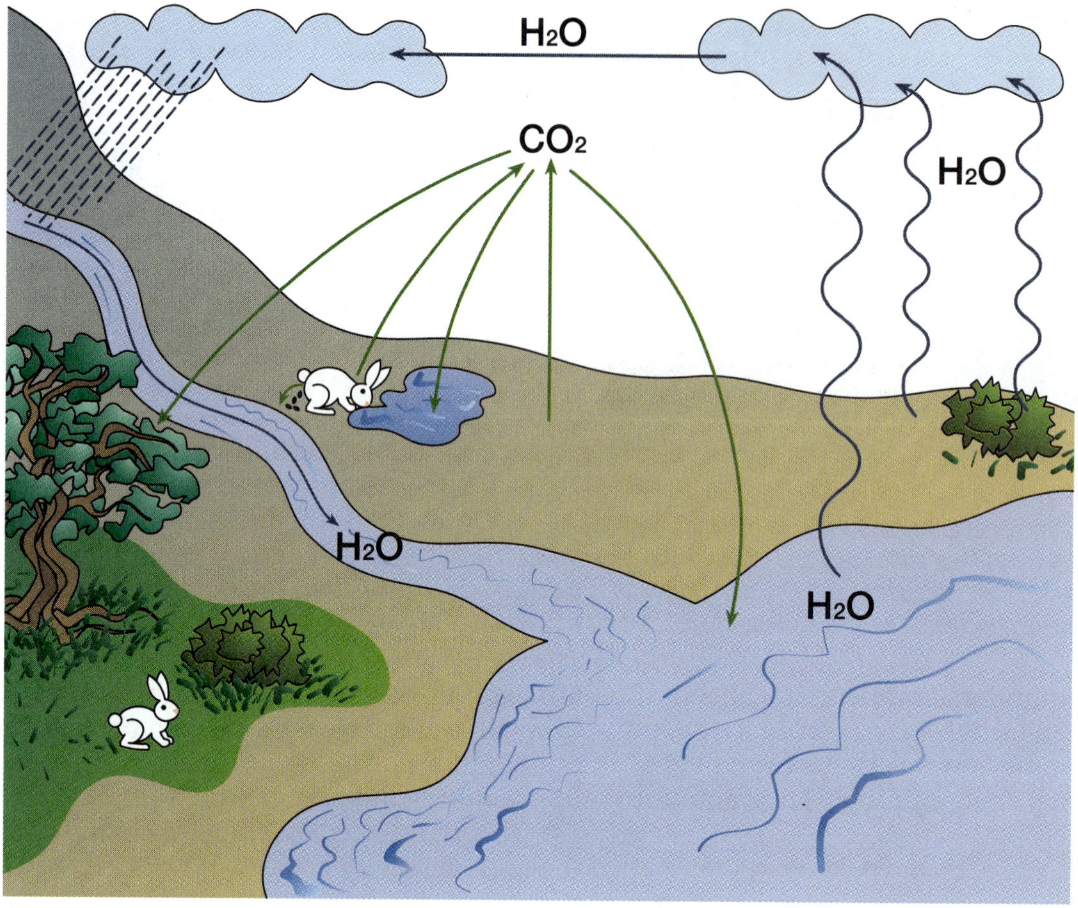

Figure 16: Carbon and water cycles *Note: Plants remove CO_2 from the air. Some of this carbon is consumed by animals. Carbon returns to the atmosphere as CO_2 when it is exhaled by living organisms and when it is released during the decay of wastes and dead material. Water evaporates mostly from soil, plant leaves, and the surfaces of oceans, lakes, and rivers. It returns to the surface as rain and snow.*

plants that remove it. The growth of the human population has also led to a great rise in the use of fuels, including wood, coal, oil, and natural gas. Fuels are burned to heat homes, cook food, move vehicles, and produce electricity. Whenever a fuel is burned, the carbon that it contains is converted to CO_2. All these activities have greatly increased the amount of carbon dioxide in the atmosphere, slowly warming the earth.

Things Are Heating Up

Earth's year-round average temperature has risen about 1°C during the past hundred years. A one-degree rise may seem small, but it is an unusually large increase in comparison with the temperature changes that have occurred in the past several thousand years. If temperatures rise another 3°C, the polar ice caps will begin to melt. Water from those huge glaciers will raise sea levels and flood coastal regions. New York, Miami, and Los Angeles might be swamped. The salt and other minerals in the oceans would be diluted by the fresh water from the ice caps. Many ocean species would probably be unable to adapt to the new conditions. They could face extinction.

The earth's rising temperature is the result of a change in how carbon recycles between living things and the atmosphere (see Figure 16). Carbon is one of many substances that are recycled. Water is another. Rainwater falls on the earth. Some of it soaks into the ground, and some flows into lakes or rivers. Plants absorb groundwater through their roots. Some of the water in plants evaporates from their leaves, and some is used in photosynthesis. A cow or other animal may eat the plant. The animal then stores the plant's water in its body or releases it as perspiration or waste. Water from plants, animals, and many other sources enters the atmosphere as water vapor (a gas). Later this water completes the cycle when it returns to the surface as rain or snow.

Every organism plays a part in these cycles. They are all linked in the great web of life. Even bacteria play essential roles. Some bacteria help to break down dead material, releasing nutrients that plants and other organisms can use. Cyanobacteria in lakes and oceans capture solar energy through photosynthesis. This energy enters the food chain when small animals eat the cyanobacteria. Other bacteria live inside organisms and trade resources with them. For example, bacteria that inhabit your intestines consume some of your food, but they also help you digest it.

A Hole in the Sky

Is your refrigerator making a hole in the atmosphere?

For years, chemicals called chlorofluorocarbons (CFCs) were widely used in refrigerators, air conditioners, spray cans, and foam packaging. These products made life easier for people. But by the 1970s the use of CFCs was beginning to look like a bad idea.

CFCs are heavier than air, but they do eventually rise to the upper atmosphere, which includes a layer of a gas called ozone (O_3). The sun's ultraviolet radiation causes CFCs to break apart and to react with the ozone. The reaction destroys the ozone at a faster rate than it can be replaced.

The ozone layer protects life from the sun's ultraviolet radiation. Damage to the ozone layer can cause skin cancer and eye disease. It can also harm other life-forms. In 1978 the United States banned aerosol sprays that contain CFCs. Steps to phase out CFCs around the globe increased after 1985, when a large hole was found in the ozone layer above Antarctica. In 1996 the production of CFCs in the United States became illegal. Today, in a worldwide partnership to protect the planet, more than 150 countries are trying to restore the ozone layer and find alternatives to CFCs.

Forming a Community

Because organisms need one another, they tend to form communities. Any community of animals, plants, and other organisms that are closely connected with one another and their environment is called an **ecosystem**. Some examples of ecosystems are a river valley with towns and farms, a desert, and a coral reef. From bacteria to people, every species in an ecosystem depends on every other species that lives there and on the environment. For example, tiny seafloor animals called polyps need warmth, minerals, and water to grow. Their skeletons form a coral reef. Some algae need coral, while other plants and animals need the algae. The reef provides a home for many species of plants and animals. Each species is a link in a complex network of nutrient and energy cycles.

Don't Spray It

Today this network is in danger because of the growth of human population and new technology. As the population has grown, so has technology. Unfortunately, some new products cause unexpected problems. For example, for many years, substances called chlorofluorocarbons, or CFCs, were used in refrigerators, air conditioners, and spray cans. But in the early 1970s researchers discovered that CFCs react with and destroy a form of oxygen called ozone (O_3).

CFCs have destroyed part of the ozone layer, which is a thick blanket of O_3 in the upper atmosphere. The ozone layer shields the earth from ultraviolet rays emitted by the sun. If you have ever suffered from sunburn, you have seen what too much ultraviolet radiation can do. It can also cause mutations, skin and eye cancers, and blindness, and it can impede the growth of important ecosystems, such as coral reefs and rain forests. The production of CFCs is an example of how one small change can have a major effect on an entire ecosystem.

Food Chain

In an ecosystem, different kinds of organisms play different roles. Plants are called **producers** because they provide the food that other organisms need for energy. In a forest, for example, plants absorb energy from sunlight and nutrients from the soil. These get passed along the food chain from one organism to another: A chipmunk eats grass and seeds; a snake eats the chipmunk; a hawk eats the snake. The chipmunk, snake, and hawk are called **consumers**. After they die, the energy stored in their bodies is released when **decomposers** (bacteria, fungi, and other organisms) break down the dead material. This allows carbon to return to the atmosphere and other nutrients to enter the soil, where plants can again absorb them, completing the cycle.

The consumers that live in a desert can tolerate the extremely dry conditions.

Each ecosystem consists of organisms that are adapted to the local environment. Imagine how different life must be in a pond in Alaska and a swamp in Florida. In the pond, algae would be the primary producers; in the swamp, water lilies and other large plants would fill this job. Both environments contain consumers, such as fish and insects. But the fish in Florida enjoy warm waters, while those in Alaska are adapted to lower temperatures. Insects that feed on the plants in these two environments are also different. In Alaska, a polar bear that eats the fish might be at the top of the food chain. In Florida, an alligator would hold the top spot.

There's a Limit

An ecosystem can support only a certain number of producers, consumers, and decomposers. Every ecosystem has limited resources. For example, fewer kinds of

Desert organisms have adapted to hot, dry conditions in various ways. The rabbit's large ears provide extra surface to radiate heat. The oily feathers of the vulture, the waxy surface of the cactus, and the dry, scaly skin of the snake and lizard help to reduce water loss by preventing evaporation.

plants and animals live in a desert than in a tropical rain forest, where water is far more plentiful. Only a few types of plants, such as cactuses and shrubs, can find and store enough moisture to grow in a desert and serve as a food source for other organisms. The consumers that live in a desert can tolerate the extremely dry conditions. Most of these creatures are reptiles and insects. A desert is too harsh for most mammals.

If certain kinds of organisms in an ecosystem are lost, the balance may be upset. For example, sloths in a jungle depend almost entirely on leaves for their water. If the plants in this ecosystem were replaced by other types with dryer leaves, sloths would not be able to live there. The types of vegetation and animals within an ecosystem are adapted to one another, and their survival is linked.

ON YOUR OWN

Respond to the following items based on "The Web of Life."

1. Pesticides used in agriculture can wipe out large numbers of insects and can leak chemicals into groundwater. Explain how excessive use of pesticides may disrupt the balance within ecosystems.

2. Floods and erosion can wash away topsoil, the nutrient-rich upper layer of earth. How could erosion in a certain environment disrupt local chemical nutrient cycles? (Hint: Think about the carbon cycle.)

3. Farmers often clear land by cutting down trees and burning them. This form of deforestation enables the farmers to produce more crops and livestock, increasing the food resources available for people. How can this land-clearing process also decrease food resources?

4. Trash buried in landfills takes many years to decay. On the other hand, some kinds of trash can be quickly burned in large incinerators. Would the use of more incinerators be a good solution to the problem of waste disposal? Explain why or why not.

LIFE IN A DEADLY PLACE

Some people haven't been home for more than 50 years. They are the former residents of the Bikini atoll in the Pacific Ocean.

In 1946 the United States tested atomic bombs on Bikini. Then in 1954 the government tested a hydrogen bomb there. Before the first tests, the U.S. government moved the human population. The islanders returned in the 1960s, but they had to leave again when scientists discovered that the islands were still dangerously radioactive.

Today sea life seems to be thriving in the waters around Bikini, but plutonium still lies at the bottom of the island's lagoon. Moreover, cesium 137, a byproduct of the atomic blasts, seeped into the soil of the islands and has contaminated its plants. To counteract this problem, scientists have been removing and replacing the radioactive topsoil. They have also experimented with potassium compounds, adding them to the soil to prevent plants from absorbing cesium 137.

In 1996 a special commission declared the islands safe enough for humans. Since then, many scuba divers have visited the site to see the ships that were deliberately sunk in the tests. The Bikinians, however, are reluctant to return to what they still consider an unsafe home. They also face another problem: Since their evacuation in 1946, their population has grown from 167 to more than 2,000.

The Web of Resources

Ever Wonder?

Do you like to eat cake or pie from the supermarket? How often do you wear jeans? How many notebooks do you have? Packaged foods, blue jeans, and notebooks are common items, but did you ever wonder where all the materials that were used to make these things came from? Does your use of these products help determine which kinds of organisms live on this planet? It may surprise you to learn that you yourself affect the web of life, the balance of resources on this planet. This investigation will help you discover how the use of a single product can affect many ecosystems and organisms throughout the world.

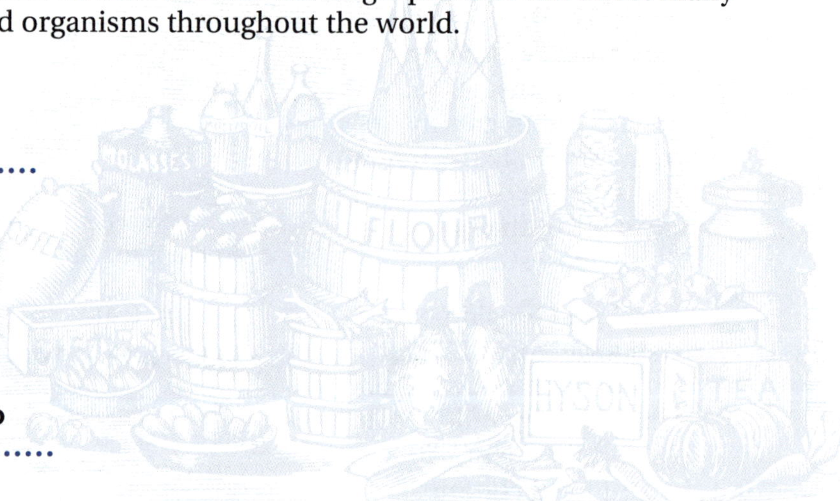

Materials List

- Miscellaneous products
- Blank map of the world
- Colored markers

Step-by-Step

1. Using the information that your group collected, make a list of places where each material in your product may have originated. (You should have two places for each material.)
2. Make a list of places where each material may have been processed, assembled, or distributed. (Again, list two or three places for each operation involved in manufacturing and distributing the product.)
3. Make a dot on your map to show the location of each place on the list you made in Step 1. Use a different color for each material or operation. (If you do not have enough markers to use a different color for each item, you may make round dots for materials and square dots for operations—or devise your own system.)
4. In your logbook, list the materials and operations in the order they occur. (For example, wheat is first grown, then ground into flour, and so on.)
5. Use colored markers to make arrows connecting the locations you marked on the map. Number these arrows in the order they occur. (Follow the list you made in Step 4.)
6. In your logbook, make a list of all the living things and natural resources that the production of your item would affect at each place on your map.

Talk It Over

Work with the other members of your group to answer the following questions.

1. How many materials are needed to make the product you researched? How are these materials obtained?
2. Do any of the materials come from farms? How would producing these materials affect the environment around the farms?
3. Which of the materials come from other countries? Which come from other continents? How would making and transporting these materials affect the environment?
4. What about the packaging of your product? Where did it come from? How does the production of the packaging material affect the environment? (If the product is not sold in a package, then consider how it is assembled and shipped. How do these operations affect the environment?)
5. What happens to the product and its package when you are finished with them? What effect does this have on the environment?
6. Many industries rely on oil, natural gas, or coal to make household products and to ship them to stores. Oil, natural gas, and coal form naturally deep in the earth, but that process takes millions of years. What will happen as these energy sources are used more and more?

Career Links

Agriculture & Natural Resources

U.S. Forest Service

Park Ranger

The main responsibilities of park rangers are to educate visitors and to protect both the visitors and the environment. Rangers must understand the needs of wildlife within their areas. They protect the parks and wildlife from vandalism, misuse, pollution, and forest fires. They also design and conduct educational talks, programs, and tours. All rangers must also be experts in emergency medical techniques and search-and-rescue procedures. Park rangers may specialize in management of natural resources, public relations, law enforcement, and other areas.

To become a ranger, it helps to have a college degree in forestry, earth science, biology, or a related field. For more information, write:

The National Park Foundation
1101 17th Street, N.W., Suite 1102
Washington, DC 20036
http://www.nps.gov/npf

Related occupations: conservationist, natural resource manager, park landscaper, botanist

Spread the Word

Work with the other students in your group to share the results of your investigation with the rest of the class. Start by telling where the materials in your product came from. Describe how the materials were originally obtained. Then discuss the various operations that combined the raw materials into a finished product. Also tell how the production of this item may have affected the environment—and how it might affect supplies of various resources in the future.

Suppose you worked for an environmental testing firm and had to write a report on the environmental effects of some product. Explain how you would research these effects. Also tell how the manufacturing of this product might affect the environment.

Read All about It!

The Disappearing Forest

THINK about all the products you see on the shelves in a supermarket: canned fruits and vegetables, jars of spaghetti sauce and salad dressing, jugs of milk, cartons of orange juice, boxes of cereal, bags of dog food. . . . Every food in the store contains ingredients that came from crops or animals that use the land. Every item contains resources from somewhere else (see Figure 15). For every snack cake that you eat, a farm must produce eggs, milk, and wheat and a factory must produce a box or wrapper. Then a truck must transport the cake to the store. During this manufacturing and distribution process, some ecosystems lose resources to others.

THE TRAIL-MIX TRAIL

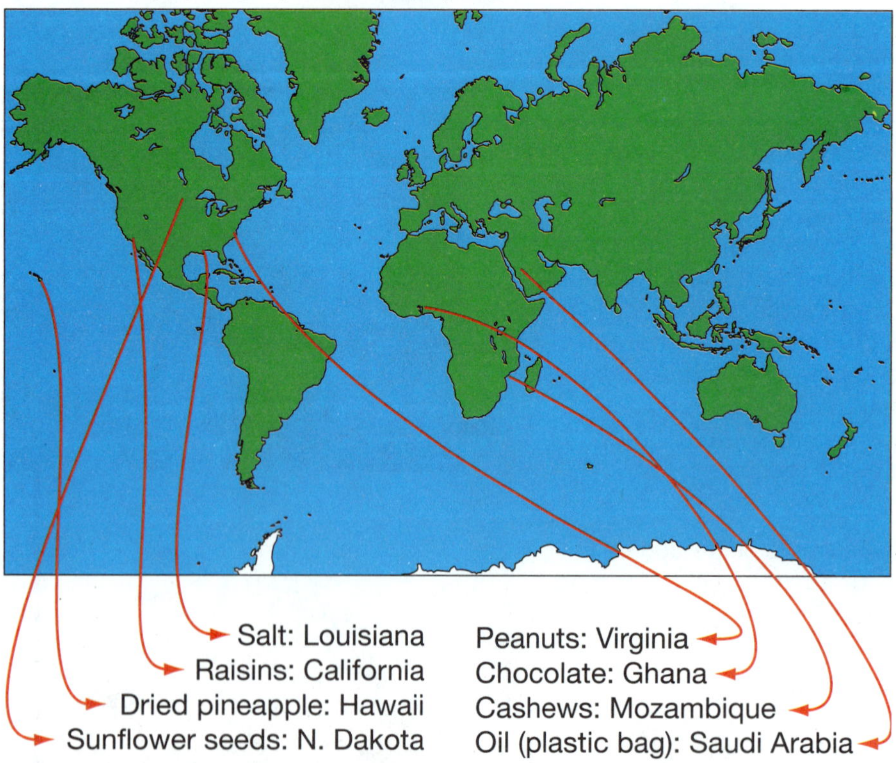

Figure 15: Origins of trail-mix ingredients
Note: Food and medicine that you purchase at local stores contain ingredients from all over the world.

Your logbook contains paper and has a cardboard cover. Where do the paper and cardboard come from? They are made from trees. Your pencil contains wood, and so does your desk—more trees. Then there are houses and furniture and fences. . . . Wood is used everywhere. To supply all the wood and paper that people need, trees must be cut.

Shrinking Forests

Forest managers have developed logging systems that allow some trees to remain standing when other trees in a forest are cut down. The trees that are not harvested provide seeds for new growth, and they maintain a habitat for wildlife while the seeds take root and develop. Foresters also plant seedlings to replace trees that have been harvested. But if such practices are not followed—if all the trees in a forest are cut down at once and if they are not replaced in some way—then the forest will disappear. The complete removal of large forests is known as **deforestation**. This kind of human activity is reducing the amount of forestland around the world.

Forests are cut not only for wood but also to make way for growing human populations and industries. Trees are a renewable resource: New trees can be grown to replace those that are cut down. Removing some trees from an ecosystem will not permanently destroy it. But if trees are not replaced or allowed to grow back, then the ecosystem cannot recover.

As the human population grows, so does its need for resources. People need homes, furniture, food, fuel, and other things. Their growing need for resources has changed ecosystems all over the world. Many tropical rain forests have been replaced by farms. To produce the coal and other minerals that people need, land is often strip-mined. Again forests are cleared. After all the minerals have been removed, the land may be left in a barren state. (Some mining companies restore the land by replanting the mined areas when they are finished.) Swamps are often drained to make room for buildings and farms. If an entire ecosystem disappears, the plants, animals, and other organisms that depend on it may die out. Some species may become extinct.

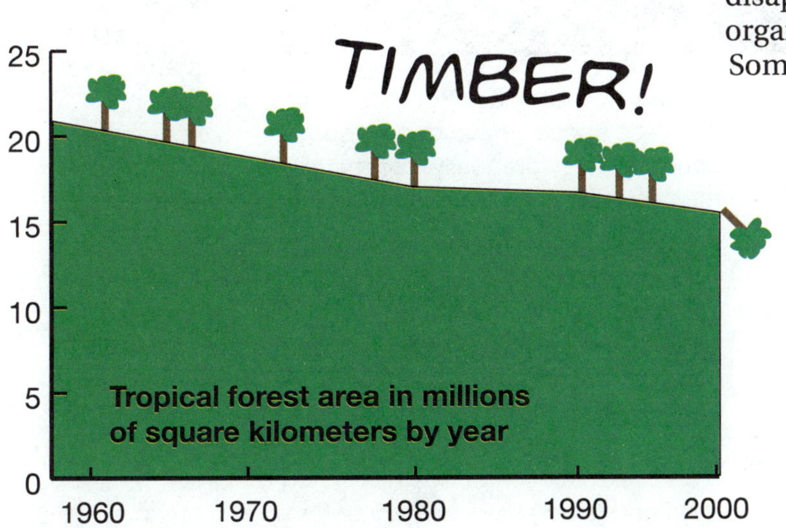

Figure 16: The disappearing tropical forest
Note: Land clearing is rapidly destroying tropical forests. In the industrial countries of North America and Europe, forest area has increased slightly since improvements were made in agricultural methods between 1910 and 1930.

The Need for Diversity

In a forest ecosystem, many species live together and depend on one another. Deforestation reduces the variety of organisms. For example, if a farmer cuts down a forest to make room for crops, the only species living there will be the crops themselves, some weeds, and the insects and microorganisms that depend on these crops and weeds. The number of species in the area—its **biodiversity**—will decline.

Without enough diversity, the new ecosystem may not do well. For example, suppose a rain forest is cleared to make room for cacao trees, a valuable crop that is used to make chocolate. With fewer species of plants to eat, the birds, insects, and other organisms that thrived in the forest may disappear. This decline in biodiversity even hurts the cacao trees. Here's why: With fewer kinds of plants in the area, tiny insects called midges disappear. Midges pollinate cacao trees, but these insects also need other kinds of plants to complete their life cycle. Without midges, the trees produce fewer cacao beans. In addition, when only a small number of crop species are planted in a large region, an insect or fungus that preys on one of those species can quickly spread, destroying the crop like an epidemic in a crowded city.

74 Species Per Day

Tropical rain forests have great diversity. In fact, the majority of the world's species live there. But the forests are being destroyed at the rate of 100,000 sq km a year—that's equivalent to clearing an area the size of a football field every second. Many species of plants and animals are becoming extinct as the forests disappear. Estimates put this number at 74 species per day.

Grounded!

What's the first thing you picture when you see the word "bird"? Is it the sight of wings stretched out in graceful, gliding flight?

Most birds can fly, but some can't. The southern hemisphere is home to a number of winged but flightless creatures. Kiwis, emus, ostriches, and cassowaries (shown here) live in Australia and New Zealand, and penguins live mostly in Antarctica. Penguins have wings that are more like flippers—they use them to swim, not to fly.

Fossil remains, such as those of the two-meter-high *Diatryma*, suggest that flightless birds appeared millions of years ago. But most, if not all, of today's flightless species have descended from ancestors that could fly. They gradually lost the ability to fly because the circumstances in which they lived did not require it.

Many nonflying birds became extinct after their environment changed. The dodo is the most famous example. This bird lived on the island of Mauritius, where it was isolated and had no predators. When sailors landed there in 1598, the dodos, who had never seen humans, did not know they were in danger. Without fear and without flight, dodos were easily caught and eaten. Moreover, because dodos nested on the ground, their eggs and young were easy prey for dogs, cats, pigs, and rats that came with the humans. By 1681 this bird, which had adapted so well to its environment, was gone forever.

These disappearing species are valuable resources. For example, medical research on some plants and animals that live in tropical rain forests has led to important drugs for heart disease and cancer. One-fourth of all prescription drugs come from plants. Researchers estimate that rain forests contain millions of species that have not yet been discovered. As the environment of these organisms shrinks, many of them may become extinct before people ever get to know them and learn their secrets.

On Your Own

Answer the following questions based on "The Disappearing Forest."

1. Widespread fires and ice ages have disrupted large ecosystems in the past. Those changes had important effects on the resources that were available to organisms. How do you think the changes affected the development of species?

2. One factor that may have been involved in the extinction of the dinosaurs was their enormous appetite. They may have consumed more food than their environment could provide. Does human population growth threaten ecosystems in a similar way? Explain your ideas.

3. To grow coffee, people in the northern and central parts of South America have cleared large areas of tropical rain forest. Would reduction in coffee drinking reduce the destruction of the forest? Explain.

4. The agave plant in the Mexican desert provides food for grazing animals and nectar for bees, insects, and birds. A species of bat depends on agave flowers for its survival, and agave depends on the bat for its fertilization. As farmland expands, the agave and the bats are disappearing. How does this situation resemble the problem with cacao plants and chocolate?

5. Most of the species that have lived on earth have become extinct. Given this fact, do you think people should care about the loss of species from deforestation? Explain your views.

Get This . . .

Average temperatures from year to year have tended to rise during the last century. This trend is called global warming, or the **greenhouse effect.**

Section Wrap-up

The Global Ecosystem

An ecosystem is a lot like your body. A pond or a forest "breathes" as the living things within it exchange gases with the atmosphere. It obtains nourishment through photosynthesis and from mineral nutrients that plants and other organisms extract from soil and water. These nutrients are "digested" as plants grow and as animals consume the plants. New generations emerge—the system reproduces. Dead material is broken down by decomposers, such as bacteria, fungi, and worms, just as your body breaks down dead cells and waste material in your blood. Wastes are recycled or passed out into the surrounding environment. The health of your body depends on the health of each of its parts. Similarly, the health of an ecosystem depends on the health and survival of each of its species.

In every ecosystem, a delicate balance exists between producers and consumers, prey and predators.

On a larger scale, you can think of the entire earth as one big ecosystem. Like your body, this global ecosystem performs many different functions. For example, the plants, algae, and cyanobacteria of the rain forests and oceans can be considered the mouth and lungs of the world—they perform much of the planet's photosynthesis. Swamps and marshes act like your liver and kidneys, breaking down and removing wastes. Such wetlands also accumulate carbon in the form of peat, muck, and decaying matter, just as your body fat accumulates energy reserves in the form of carbon compounds. That helps keep excess carbon dioxide out of the air.

Traffic clogs the streets in Jakarta, Indonesia. Exhaust from vehicles and industry add CO_2 and other pollutants to the atmosphere.

It All Hangs Together

What happens to one part of your body—or one part of an ecosystem—affects the other parts. If you cut a finger and it gets infected, that infection can spread throughout your body, causing serious illness. A stomach virus takes away your appetite, and that means less food and energy for your whole body. In an ecosystem, if a river floods surrounding land, some plants, insects, mice, and other small animals may drown. Birds and larger animals may have to leave the area to find food. This will increase competition for food in neighboring places. Sewage, fuels, and chemicals may contaminate the water, entering food chains in the flooded area and downstream. Finally, if contaminated floodwater reaches the sea, it dilutes the salt water in that area and delivers nutrients and pollutants that disrupt the shoreline ecosystem.

Changing the balance can threaten the entire ecosystem.

Too much of the same thing can make you sick. Too much sunlight can give you a sunburn; too much candy can lead to tooth decay; too much caffeine puts stress on your heart. An ecosystem can also suffer from too much of the same thing. If too many members of one species live there, the balance will be upset. If too much acid appears in the soil, fewer plant species will be able to survive. Too many deer can overgraze a forest, causing severe damage to trees. Too many turtles can threaten the frog population. In every ecosystem, a delicate balance exists between producers and consumers, prey and predators. There are only so many resources to go around. Changing the balance can threaten the entire ecosystem.

Demanding People

A planet also has limited resources. But the growing human population keeps demanding more food, more space, and more natural resources. The more the human race grows, the more the environment is changed to meet people's needs. For example, as more fuel is burned in cars, homes, and factories, more CO_2 and other pollutants enter the atmosphere. People are changing the way the planet "breathes."

> **Get This . . .**
> **Biodiversity** refers to the number and variety of species in an area.

Air pollution has also changed the "nutrition" that the world receives from solar energy. As atmospheric levels of CO_2, CFCs, and other gases increase, average temperatures around the world are rising and the protective ozone layer is getting thinner. If these trends continue, major changes could occur. Higher temperatures could melt the polar ice caps, causing sea levels to rise. A thinner ozone layer that allows more dangerous radiation to reach the surface could harm many people.

All over the world, forests and other natural ecosystems are losing ground to logging, farming, and housing. The more bananas, chocolate, and coffee that people consume, the more tropical land they use to grow these crops. Clearing land to grow them adds CO_2 to the atmosphere. At the same time, as wages rise in developing nations, their people are demanding more cars and other goods from industrial countries and more fuels from oil-producing ones. Because of this, air pollution increases in all three regions. Demands in one region affect ecosystems all over the world.

Who's Worried?

How will these changes affect the future of this planet? Some people point out that changes in temperature, atmosphere, and species have occurred in the past. They argue that such changes are a natural part of the earth's history and that there is no reason to be concerned. Other people worry that damage to ecosystems is occurring too rapidly. They claim that people are destroying their own future by damaging systems that all living things need.

Which group is right? You can make your own decision, but one thing is clear: The global environment has been changing rapidly for the past hundred years. No one knows what will happen during the next hundred years. But whatever changes occur, people will probably play the most important role in all of them. Changes in how they live—for example, by burning cleaner fuels and by preserving rain forests and other wild ecosystems—could make all the difference in the world.

On Your Own

Answer the following questions based on "The Global Ecosystem."

1. How does recycling alter the kinds of changes that are occurring to the global environment?
2. A sick person can sometimes improve if a surgeon replaces a diseased organ. Can global environmental problems be reduced by replacing a ruined ecosystem with a new one? Explain.
3. Many desert plants sprout or bloom at the same time, immediately after it rains. What are some possible effects that a change in climate would have on desert plants and animals?

MODULE WRAP-UP

Time Trails—A Study of the History of Life on Earth

A FIELD trip to the zoo. Everyone is glad to get outside. On the way your teacher says, "Try to apply what you have learned about natural selection and speciation while you are there."

At the zoo, a bridge takes you over a swampy area. Something smells terrible. Yuk—the murky water is full of rotting food. The odor reminds you that the first life on earth had to survive without oxygen to break down smelly substances. The oxygen in today's atmosphere would have killed the early bacteria.

A Real Survivor

You see a huge crocodile on a rock, sunning itself to keep warm. This scaly reptile is like a glimpse of the distant past. Its ancestors moved out of the water to live on land. With their powerful jaws and armor plating, it's easy to see how these creatures have survived while many other prehistoric species went extinct. Reptiles have scaly skins that keep them from drying out on land. A crocodile's open mouth allows it to release heat. Unlike reptiles, amphibian species, such as frogs, have moist skins that keep them cool.

Inside the reptile building, you are eyeball to eyeball with hundreds of species of snakes and lizards—big and small, patterned and striped, you name it. All this diversity shows how a species can change over millions of years. Reptiles developed early, so they spread to every continent and adapted to many environments. Some climb trees; some slither through weeds; others hunt in water. In some environments, reptiles are at the top of the food chain.

Back outside, the air is filled with the songs of birds in the aviary next door. It's hard to believe these soft, fragile-looking creatures evolved from reptiles, but their claws and scaly legs are evidence that they did. Their feathers correspond to scales. Wings help birds escape predators and find food and water. Reptiles can be trapped between mountains or oceans, but birds can fly over these natural barriers.

Something moves. It's a bee in a flower near your hand. In fact, many bees are swarming over the pink and yellow blossoms, searching for nectar. Bees and flowers have an excellent relationship: The flowers provide nutritious nectar for the bees, and the bees spread the pollen that fertilizes the flowers.

The ants have also hit the jackpot. Hundreds of them are carrying off bits of food. Without noticing, people are stepping on some of them. Like most insects, however, ants produce many offspring. Insects are eaten by so many other organisms that they needed to develop this adaptation in order to survive.

Dangerous and Endangered

Now it's on to the lions. Their home looks like their natural habitat in Africa: a large open area with rocks and trees, surrounded by a moat. But their natural habitat keeps shrinking, as the rapidly growing human population pushes farther into their range. It is possible that someday the only places where lions and other large wild animals live will be in zoos and game preserves.

At least the lions are doing better than the black

rhinos across from them. On the fence around the rhinoceros pen is a sign reading "Endangered." Very few of these huge mammals remain. They have been hunted

even more than lions. With so few of them left, the rhinoceros "gene pool" is more like a puddle—few genetic variations exist among the remaining rhinos. Future offspring probably will not be varied enough for the species to survive.

Bears. Even though bears have not been around as long as many other kinds of animals, many species have developed and they have adapted to a wide variety of environments. A panda, their close relative, sits nearby, using its thumb to grip and peel a bamboo shoot. No bear could do that. This adaptation helps the panda survive in Southeast Asia. A polar bear's paws are different: Wide pads and sharp claws are more suited to the Arctic, where they allow the bear to walk on ice and kill seals.

Monkeying Around

Suddenly a racket erupts on Monkey Island. A small adult monkey has tried to snatch some food from one of the babies, and a group of the other adults ganged up to chase him off.

Everyone in your group is surprised to see how monkeys cooperate. Their little island is an organized community. It's a little bit like an ecosystem: Each organism on the island depends on all the other organisms, and their delicate balance can be easily upset. It was only a little food, but its effect on one member spread to the whole group.

Human activities often disrupt ecosystems. When forests are cut and burned or when air and water are polluted, many delicate ecosystems are upset. Species that depend on these systems may not survive.

But forests cannot be protected as easily as monkey food; nor does pollution go away as quickly as a greedy monkey. As more and more people drain resources from the environment, life itself is challenged to adapt—fast.

On Your Own

Answer the following questions based on the Module Wrap-up.

1. How is the relationship between the bees and the flowers like one theory of the origin of eukaryotes?

2. The Australian emu and the African ostrich are very similar birds that cannot fly. Both species appeared after the continent of Australia broke away from Africa. How could they develop in such widely separated places?

3. What could have led the ancestors of some air-breathing animals, such as dolphins and seals, to live in the sea? Explain your ideas.

4. In some reptile species, including crocodiles, the sex of the young depends partly on the temperature at which the eggs develop. How could deforestation and the greenhouse effect influence the future survival of these species?

How to Use Your Logbook

All through *Science Links* you will use a logbook to keep records of your work. A logbook is an important tool in scientific investigation. It provides a place to take notes, collect data, enter observations and conclusions, and jot down questions and ideas. Most scientists keep a log like this one.

In science it is essential to keep accurate records. The information that you collect is valuable—you must work to get it! So write down your ideas before you forget them.

Your logbook will also help you share your ideas with other students. In most of your investigations, you will work as a member of a group. In any situation where people must work together, whether it's flying a jetliner or leading cheers at a basketball game, teamwork is critical. One of the most important aspects of teamwork is communication. No team can succeed unless its members communicate effectively with one another.

Good Writing Is Clear and Simple

Effective communication means writing and speaking in ways that other people will understand. Your logbook can help you develop a good writing style. Whenever you make an entry, try to express your thoughts as clearly and simply as possible. When you finish writing, step back and look at the words as though you are reading them for the first time. If someone else had made this entry, would you be able to understand it? Is the writing clear? Does it cover all the points it was supposed to? You know what you meant to say, but did you succeed in saying it? If you have any doubts, go back to work on the entry.

Writing makes you more aware of your thinking process. This is useful in science because sometimes you don't even realize that you do not understand a concept until you try to express it. Communicating scientific ideas through writing and discussion helps you understand what you are doing. And in *Science Links* the emphasis is on doing science.

Your logbook gives you an opportunity to show your teacher what you have learned. It will reveal how you and the other members of your group are collaborating. The notes you take during presentations by other groups will show if you are receiving useful information from them.

Organizing Your Logbook

Your logbook should be organized chronologically. Every time you make a new entry, put the date in the upper left-hand corner of the page. Next, in the upper right-hand corner, write a few words that tell what the entry is about. Then write the entry itself. (A sample entry appears on the next page.)

Various types of entries are possible. Here are some of them:

- notes based on what occurs during procedures **(Step-by-Step)**
- summaries of group or class discussions **(Talk It Over)**
- outlines for presentations by your group **(Spread the Word)**
- notes and questions about presentations by other groups
- data collected through research
- ideas for projects or reports
- personal views concerning *Science Links* activities

Use your logbook regularly. It will help you keep track of and understand the science you are doing in *Science Links*.

September 30, 1998

Module 2, Investigation 4
Calculations and Observations

Step-by-Step

Step 4	baking soda & paper	5.46g
	paper	−3.46
	baking soda	2.00g
Step 9	beaker & contents	123.76g
Step 12	beaker & contents after reaction	−122.31
	weight lost	1.45g

After Kim poured the vinegar in the beaker, it foamed up. Giselle said it smelled sour.

Step 14	baking soda & paper	4.83g
	paper	−2.83
	baking soda	2.00g
Step 21	bottle & contents	90.50g
Step 23	beaker & contents after reaction	−90.30
	weight lost	0.20g

After I poured the vinegar out, it foamed up.

Talk It Over

#1–4. Everybody agreed that the beaker lost more weight than the bottle. Kim said this was because something turned into a gas and it went out of the beaker but couldn't get out of the bottle.

Module Glossary

adapt—to change in a way that increases an organism's ability to survive and reproduce in the environment (6)

adaptive radiation—separation of a population into species adapted to survive in different environments or to use different resources in the same environment; this process reduces competition within the population (14)

aerobes—organisms that require oxygen to survive (27)

amphibians—vertebrates, such as frogs and salamanders, that require water for reproduction and that spend part of their lives in water and part on land (38)

anaerobes—organisms that can live without oxygen; most anaerobes are bacteria for which oxygen is poisonous (26)

archaebacteria—a group of mostly anaerobic bacterial species that are thought to be similar to some of the first living cells; many of these species obtain energy by performing chemical reactions on iron, sulfur, and other minerals (21)

biodiversity—the coexistence of a number of species, especially in the same environment (73)

biogeographical realm—a geographical region populated by particular species that are not found in similar environments outside the region; these population patterns are affected by continental drift, the age of species, and the ease with which species are able to spread beyond their places of origin (55)

competition—the relationship between organisms or species that use the same resources; competition reduces the survival and reproduction of the individuals involved (5)

consumer—an organism that obtains nutrients from other living things (68)

continental drift—the slow movement of the earth's crustal plates that carry the continents (54)

convergence—adaptation of unrelated species, such as fish and dolphins, to similar environments, with the result that they come to resemble one another (56)

cyanobacteria—small, bluish-green, photosynthetic bacteria (27)

decomposer—an organism, such as an earthworm or fungus, that obtains nutrition from dead material or the wastes of other organisms (68)

deforestation—the removal of trees and the replacement of forestland with another type of ecosystem, such as farmland or grassland (73)

deoxyribonucleic acid (DNA)—the substance that carries genetic information in all living cells (20)

ecosystem—a community of organisms that depend on one another; this kind of community usually includes a number of different types of producers, consumers, and decomposers (67)

endosymbiont hypothesis—the theory that eukaryotic cells originated when several types of prokaryotic cells either joined together or were ingested by other prokaryotic cells (28)

eukaryotes—organisms, such as plants, animals, and fungi, whose cells contain a nucleus and other organelles (28)

exoskeleton—a shell, such as those of crabs and insects, that provides support and protection to softer, internal body parts (38)

geographic isolation—a situation in which one population is separated from another population by a barrier, such as a mountain range or desert, that prevents members of the two populations from interbreeding (14)

greenhouse effect—the increase in air temperature that occurs when an atmosphere containing carbon dioxide is exposed to light; the greater the CO_2 concentration is, the more rapidly the temperature increases (65)

hydrothermal vents—cracks in the ocean floor in which seawater is heated to boiling by the underlying magma (21)

marsupial—a type of primitive mammal, such as the kangaroo and the koala, in which immature young develop in a pouch on the mother's abdomen (55)

mutation—any change in an organism's hereditary material (13)

natural selection—an improvement in a population's adaptation to its environment because of the greater survival and reproduction of the better-adapted members of the population (6)

Pangaea—a single landmass made up of all the earth's continents; Pangaea is thought to have existed during the Paleozoic and Mesozoic eras, which ended about 200 million years ago (43)

photosynthesis—the process whereby plants, cyanobacteria, algae, and a few other types of organisms form sugars from carbon dioxide and water (27)

placental—a type of mammal in which the young complete their embryonic development within the mother's womb (55)

polymers—large molecules, such as starch, protein, and DNA, that are made up of many similar small molecules (21)

producer—an organism that uses a nonliving energy source, such as sunlight, to produce new living material (68)

prokaryotes—bacteria; simple-celled organisms without a nucleus (26)

ribonucleic acid (RNA)—a substance similar to DNA that occurs in all living things and is important in the expression of the hereditary information in DNA; RNA is also the hereditary material in some viruses; according to one theory, RNA may have originated before DNA or protein (20)

selection pressure—the effect of the environment in natural selection; it favors the survival and reproduction of some types of individuals over others in a diverse population (6)

speciation—the separation of a population over time into species that can no longer breed with one another (14)

species—a population of individuals capable of interbreeding and producing fertile offspring (4)

variation—a difference among individuals in a population (4)

vertebrates—any of the five types of animals with backbones: fish, amphibians, reptiles, birds, and mammals (50)

For Further Study . . .

Books and Articles

Dixon, D. 1990. *Man after Man: An Anthropology of the Future.* New York: St. Martin's.

Friday, G. 1997. Environmental niches. *Science Teacher* 64 (May): 38–41.

Gould, S. 1992. *The Panda's Thumb: More Reflections in Natural History.* New York: Norton.

———. 1989. *Wonderful Life: The Burgess Shale and the Nature of History.* New York: Norton.

Karl, T., N. Nicholls, and J. Gregory. 1997. The coming climate. *Scientific American* 276 (May): 78–83.

Madigan, M., and B. Marrs. 1997. Extremophiles. *Scientific American* 276 (April): 82–87.

Nardo, D. 1995. *Dinosaurs: Unearthing the Secrets of Ancient Beasts.* San Diego: Lucent Books.

Pennisi, E. 1996. Beetle ears befuddle bats. *Science* 271: 148.

Pinter, N., and M. Brandon. 1997. How erosion builds mountains. *Scientific American* 276 (April): 74–79.

Rice, R., R. Gullison, and J. Reid. 1997. Can sustainable management save tropical forests? *Scientific American* 276 (April): 44–49.

Tattersall, I. 1997. Out of Africa again . . . and again? *Scientific American* 276 (April): 60–67.

Turco, R. 1996. *Earth under Siege: From Air Pollution to Global Change.* New York: Oxford University Press.

Wilson, E. 1993. *The Diversity of Life.* New York: Norton.

Internet

Agency for Instructional Technology (AIT):

- World Wide Web—http://www.ait.net
- *Technos,* a journal for education and technology—technos@ait.net

South-Western Science:

- World Wide Web—http://science.swep.com
- Science Discussion List—To join, send an e-mail message to majordomo@list.thomson.com and in the body of the message include the words *subscribe* swep-science.

Video

The Peppered Moth: A Population Study. 1986–87. This video examines the connection between environmental changes and variations in a population of moths. Available from Media Design Associates, Inc., P.O. Box 3189, Boulder, CO 80307-3189.

The following video programs are available from the Video Project, 200 States Drive, Ben Lomand, CA 95005.

From the Beginning. 1995. This 45-minute video explores life on earth from the earliest living things through humans, addresses theories of evolution, and traces human development from embryo to birth.

Garbage into Gold. 1995. The benefits of recycling are presented by portraying real-life community and business solutions to the problem of solid-waste disposal.

Greenplans. 1995. This program explores using consensus building to solve severe environmental problems of pollution, land use, and resource development in the Netherlands and New Zealand.

The following three video programs are available from AIT, Box A, Bloomington, IN 47402.

Pollution. 1992. This program explains how humans pollute the environment so much that natural biological cycles cannot absorb all the waste. It also introduces ways to prevent pollution. A Science Source video. 15 minutes.

Climate. 1990. This video explores climatic changes—including the greenhouse effect—and their consequences for human activity. It contains an optional computer component. From the Interactions in Science and Society series. 30 minutes.

Life on Our Planet. 1992. This series of six 15-minute videotapes investigates the scientific concepts behind the earth's biological forms. Individual programs focus on the diversity of life, bees, insects, plants, tropical regions, and water and its life systems. The programs are available separately or as a series.

Software

Evolution. 1995–96. This CD-ROM features such topics as natural selection, sources of variety, and speciation. The software is compatible with Macintosh, Windows, or DOS. It is available from Media Design Associates, Inc., P.O. Box 3189, Boulder, CO 80307.

The following software packages are available from Maxis, Education Team, 2121 N. California Blvd., Suite 600, Walnut Creek, CA 94596-3572.

SimIsle. The user rules a tropical island and must make choices that will affect its future. Poachers, endangered species, tourists, developers, and movie crews are part of the cast.

SimLife. In this computer simulation, the user learns about life by designing an entire ecosystem from scratch.

SimEarth. This software allows the learner to design simulated planets and their life-forms.

All three packages are compatible with MS-DOS (3.3 or higher), Macintosh (System 6 or higher), and Windows (3.0 or higher).

Agencies and Organizations

Environmental Protection Agency. The EPA's mission is to protect public health and to safeguard and improve the natural environment. Its web site contains a page for students and teachers, with links to dozens of appropriate publications and curriculum items. Contact the U.S. Environmental Protection Agency (EPA), Environmental Education Division, 401 M Street, S.W. (1707), Washington, DC 20460; web site: http://www. epa.gov.

National Oceanic and Atmospheric Administration. The mission of the NOAA includes describing and predicting changes in the global environment. The address for its National Climatic Data Center is Federal Building, 151 Patton Avenue, Asheville NC 28801-5001; web site: http://www.ncdc. noaa.gov.

U.S. Global Change Research Information Office. The GCRIO provides information on global-change research, strategies and technologies, and educational resources. Contact GCRIO User Services, 2250 Pierce Road, University Center, MI 48710; web site: http://www.gcrio.org.

Illustrations and Photos

Illustration Credits—Mike Cagle, 9, 21, 28 (right), 39, 50, 54, 58 (top), 62, 64, 66; Corel Corporation, 43, 58 (bottom); Brenda Grannan, icons; Jay Hagenow, 27, 44, 45, 46 (time line), 55, 65 (graphs), 72, 73; Vance Lawry, 28 (left), 29, 44; T/Maker Company, 47

Photo Credits—Corel Corporation, 4, 6, 7, 13, 14, 15, 16, 19, 30, 38, 52, 56, 60, 65, 67, 68, 73, 74, 76, 78, 79; Jay Hagenow, 18, 34, 57; Dr. Mark R. Meyer, 3 Dot Studio, viii–1; NASA/Corel, 25; National Park Service, 26; Dr. Geoff Oxford, University of York, 5